This textbook is designed to fulfill the mathematics curriculum requirements for community elementary schools. It is being proposed and recommended as a mathematics text especially for use by pupils in the sixth grade in English-speaking countries.

It contains valuable information to help pupils understand fundamental principles and concepts, integrated with numerous problems with answers, of Elementary Mathematics. Other titles in J. Nyenetu Jarkloh's series of Elementary Mathematics are:

Modern Mathematics for Community Elementary Schools (Grade 5)

Modern Mathematics for Community Elementary Schools (Grade 5): Teacher's Edition

Principles of Modern Elementary Mathematics (Grade 6): Teacher's Edition

PRINCIPLES OF MODERN ELEMENTARY MATHEMATICS

GRADE 6

J. NYENETU JARKLOH

Paperback Edition

Copyright © 2013 by Nyenetu Jarkloh

Printed by: CreateSpace™, an Amazon.com Company
Available from Amazon.com and other retail outlets, online stores as well
as book stores and on Kindle and other devices.

ISBN-13: 978-1490319988 (CreateSpace Assigned)
ISBN-10: 1490319980

For information on how to order this book or other titles of the author's
series of Elementary Mathematics, please contact:

https://www.amazon.com/author/nyenetujarkloh
https://www.createspace.com/4306021
E-mail: nyenetujarkloh2010@yahoo.com

DEDICATION

This book is dedicated to
Ella, the love of my life, for her moral support and being a dedicated wife
and loving mother;
And to Julian, my son, for his encouragement and graphic design
support in drawing some of the figures in this book.

Table of Contents

PRINCIPLES OF MODERN ELEMENTARY MATHEMATICS
GRADE 6

ACKNOWLEDGEMENTS

The writing of a book is a tedious and complex undertaking. I would like to express my thanks to all those who, during the preparation of this book, have contributed to it in no small measure. In particular, I acknowledge the contributions of Dmitry Ilich and Alexandr Yevmenchikov of Orion Group of companies in drawing some of the geometric figures.

I acknowledge the contribution of Dr. Vyacheslav S. Shobik, Associate Professor and Chief of Foreign Affairs Department of Odessa National Polytechnic University. Dr. Shobik had reviewed and pointed out some mistakes contained in the original text of the book; and, despite his busy time at work, he was always available when I needed to consult with him. I am sincerely grateful for his editorial and moral support.

I am also very grateful to Evgeniy N. Bogdanov, owner of the Sevastopol-based company called "BrainCrackers.com". He was helpful in converting the MS Word interior file of the book into the PDF format required for publication.

My special thanks go to my son Julian Jarkloh for also drawing some of the geometric figures, suggesting and doing the design of the front cover.

My friend Nikolay Arminevich Danilchenko of TransSibGas accorded me the hospitality to print out the finished text of the book in preparation for publication. I am thankful for his usual material assistance, which was most valuable in getting the book ready for publication.

Finally, this book would never have been completed without the moral support of my wife Ella, who always believed in me even in those difficult circumstances when I felt that it wasn't possible. I am sincerely grateful for her support, patience and encouragement while I wrote this book.

Foreword

"Principles of Modern Elementary Mathematics (Grade 6)" is a sequel to "Modern Mathematics for Community Elementary Schools (Grade 5)". It is divided into four chapters, and does not contain the accompanying manual "Students' Mathematics Workbook for Grade 6", which contains solutions and explanations to problems and exercises. Nevertheless, it cherishes the traditional curriculum objective to build a solid foundation for pupils so that they can acquire the aptitude required to successfully continue the study of mathematics in upper classes. Handbook materials are included in the appendix for easy reference for both pupils and the teacher.

Although it does not contain the "Students' Mathematics Workbook for Grade 6", the text is integrated with ample example problems and their solutions to assist and guide pupils in understanding the practical applications of theoretical concepts. Besides, answers to "Problems and Exercises" at the end of each chapter are provided at the end of the book. Pupils attempting to solve these "Problems and Excersies" at the end of each chapter would find this feature very useful. In general, the book is highly recommended as an invaluable companion to help pupils in getting the best possible experience and results in the study of 6th grade mathematics.

The following is a description of the book's content.

Chapter one presents a comprehensive coverage of the principles of elementary mathematics, focusing on the fundamental concepts about percentages, ratios and proportions, simple and compound interests, as well as the properties of direct and indirect proportionalities.

Chapter two gives additional treatment on elementary geometry, integrated with extended definitions of geometric terms, to build on relevant geometric concepts already acquired in grade 5. Parallel and perpendicular lines are examined; the different kinds of quadrilaterals are discussed, as well as areas of triangle, rectangle, parallelogram, and trapezoid, including concepts on prisms and pyramids. Chapter three deals with circular and spherical bodies. In particular, the lengths of circumferences and areas of circles, as well as concepts regarding cylinders and cones are considered from both theoretical and practical standpoints.

Part two of the book, the solution manual, contains chapters 4 through 7. Chapters 4, 5, and 6 contain complete solutions and answers (with explanatory notes) to all of the *Problems and Exercises* included at the end of chapters 1, 2, and 3 in Part one. The solutions embody the methods and examples of dealing with typical problems and exercises with different levels of difficulty contributing to the learning environment as well as promoting the mastering of the materials and concepts presented. Chapter seven is composed of five aptitude tests, which feature typical problems and exercises within the context of "Modern Mathematics for Community Elementary Schools (Grade 5) and "Principles of Modern Elementary Mathematics (Grade 6)".

It is hoped that this book will not only enable both 6th grade mathematics teachers and their pupils to more effectively use the time allotted to the study of "Principles of Modern Elementary Mathematics (Grade 6)", but also to make the teaching environment and learning process more enjoyable, more friendly and more comfortable. The theoretical platform of the present book integrated with its practical approach is a very unique innovation in the design and conceptualization of contemporary elementary text. I hope that it would be a modest contribution to the development and improvement of national and regional school curriculum on elementary mathematics.

J. Nyenetu Jarkloh.

Chapter One: Percentages, Ratios and Proportions

1.1 Percentages

1.1.1   Introduction and General Concept about Percentages

Almost all the time people are faced with the need to perform one calculation or another in offices, at homes, in shops, and practically any place where it is necessary to deal with and establish the quantitative descriptions or relationships of objects, items, and values. They use different numbers and different ways of expressing them to represent one quantity or another. For example, it is possible to say 30 minutes, or 0.5 hour, meaning one half ($\frac{1}{2}$) of an hour. Also it is possible to say 700 grams, or $\frac{7}{10}$ of a kilogram, or 0.7 kg. One millimeter can otherwise be expressed as 0.1 cm, or $\frac{1}{10}$ cm. One decimeter is equal to 10 cm, or $\frac{1}{10}$ of a meter. There are so many examples to think about.

In the same way, *a particular name is given to one hundredth of any quantity, number, or value. One hundredth is called a percent. It means, if you divide any whole quantity, number, or value into hundred parts, then each part is said to be one percent of the whole. A percent is symbolized by a special sign %.* Examples of *percents* can be written as:

$$\frac{1}{100} = 0.01 = 1\ \%; \quad \frac{25}{100} = 0.25 = 25\ \%; \quad \frac{85}{100} = 0.85 = 85\ \%.$$

The word *percent* is of a Medieval Latin origin, having the root *"cent"*, which means "hundred". Together with the preposition "per" (per centum), the word "percent" literally means *"out of every hundred"*. In Ancient Rome, money paid by a debtor to a creditor for every hundredth piece of a land was referred to as percent. Under our present-day economic system, percent may be defined as security yielding a rate of interest as specified. Security is a certificate (or the financial asset represented by such a certificate) of creditorship or property carrying the right to receive interest or dividend, such as shares or bonds.

Common percents and their corresponding common fraction and decimal fraction equivalents are the following:

(a) $5\% = \dfrac{5}{100} = \dfrac{1}{20} = 0.05;$

(b) $10\% = \dfrac{10}{100} = \dfrac{1}{10} = 0.10;$

(c) $20\% = \dfrac{20}{100} = \dfrac{1}{5} = 0.20;$

(d) $25\% = \dfrac{25}{100} = \dfrac{1}{4} = 0.25;$

(e) $40\% = \dfrac{40}{100} = \dfrac{2}{5} = 0.40;$

(f) $50\% = \dfrac{50}{100} = \dfrac{1}{2} = 0.50;$

(g) $60\% = \dfrac{60}{100} = \dfrac{3}{5} = 0.60;$

(h) $75\% = \dfrac{75}{100} = \dfrac{3}{4} = 0.75;$

(i) $80\% = \dfrac{80}{100} = \dfrac{4}{5} = 0.80;$

(j) $90\% = \dfrac{90}{100} = \dfrac{9}{10} = 0.90.$

1.1.2   Finding a Percent from a Given Number

Problem. In a processing plant, 75% of sugar by mass is obtained from sugar cane. What quantity of sugar by mass can be produced from 500 kg of sugar cane?

Solution. There are different methods of solving this problem. Let us look at two of them.

First method: In order to solve this problem, it is necessary to find 75% of 500. Since 75% is equal to $\dfrac{3}{4}$ or 0.75, then the problem can be solved by finding the fraction of the given number:

$$75\% \text{ of } 500 = \frac{3}{4} \text{ of } 500 = 0.75 \text{ of } 500 \Leftrightarrow \frac{75}{100} \cdot 500 = 75 \cdot 5 = 375 \text{ (kg)}.$$

This means that 375 kg of sugar can be produced or obtained from 500 kg of sugar cane.

Second method: This problem can be solved by another method:

$$\text{We first find } 1\% \text{ or } \frac{1}{100} \text{ of } 500 \Rightarrow \frac{1}{100} \cdot 500 = \frac{500}{100} = 5 \text{ kg};$$

In order to find 75% of 500, it is necessary to multiply: $5 \cdot 75 = 375$ (kg).

This method of solution to the problem may be written in the form of a numerical expression:

$$(500 \div 100) \cdot 75 = \frac{500}{100} \cdot 75 = 5 \cdot 75 = 375 \text{ kg (answer)}.$$

1.1.3   Finding a Number by a Given Percent

Problem 1. What quantity of sugar cane can be processed under a defined set of conditions in order to obtain or produce 375 kg of sugar, if the quantity of sugar makes up 75% of the total mass of sugar cane?

Solution. This is a reverse problem to the preceding one. Let us look at three methods of solving this problem.

First method: 75% of sugar can be obtained from sugar cane; that is, 375 kg of sugar constitutes 0.75 of the total mass of sugar cane. The nature of the problem is finding the number by its given percent, i.e. finding the total mass of sugar cane that can produce 375 kg of sugar at the given percent:

$$\frac{375}{0.75} = \frac{37500}{75} = 500 \text{ (kg)};$$

This means that 500 kg of sugar cane can be processed (under a defined set of conditions) to produce 375 kg of sugar.

Second method: Since 375 kg of sugar makes up 75%, then the quantity corresponding to one percent would be: $\frac{375}{75} = 5$ (kg). All the sugar cane is considered as a whole unit, i.e. it is 100%. Therefore, in order to determine the total mass of sugar cane, it is necessary to multiply by 100: $5 \cdot 100 = 500$ (kg).

The solution to the problem may be written in the form of a numerical expression:

$$\frac{375}{75} \cdot 100 = 500 \text{ (kg)}.$$

Third method: Let x be equal to the quantity of sugar cane from which it is possible to obtain 75% of sugar. Then the expression ($\frac{x}{100} \cdot 75$) represents the quantity of sugar produced or obtained from the total mass of sugar cane.

According to the problem, the said quantity of sugar produced or obtained is equal to 375 kg. Therefore, we write and solve the equation:

$$\frac{x}{100} \cdot 75 = 375;$$
$$\frac{x}{100} = \frac{375}{75};$$
$$x = \left(\frac{375}{75}\right) \cdot 100;$$
$$x = 500 \text{ kg.}$$

It is left with the pupil to judge and apply the method of solution to the problem, which she or he considers more convenient.

Problem2. Mango contains 65 % of nectar; and nectar contains 90 % of honey. What quantity of mango is necessary in order to obtain 25 kg of honey?
Solution. Two methods are suggested below.

First Method:
(a) According to the problem, we can obtain 90 kg of honey from 100 kg of nectar. Therefore, we need to find the quantity of nectar necessary to obtain 25 kg of honey. So we represent that quantity of nectar by x, and reduce this into equation form:

100 kg of nectar = 90 kg of honey,
x (kg) of nectar = 25 kg of honey.

In other words, 100 kg of nectar is to 90 kg of honey as x (kg) of nectar is to 25 kg of honey:

$$\frac{100}{90} = \frac{x}{25};$$

Cross multiplying, we have the following:

$$90x = 25 \cdot 100 \Rightarrow x = 2500 \div 90$$

$x = 27.78$ (kg) – the quantity of nectar necessary in order to obtain 25 kg of honey.

(b) We follow similar approach, which is based on the data given in the problem and the quantity of nectar needed to obtain 25 kg of honey (27.78 kg - as we have already determined above). The data given in the problem is that mango contains 65 % of nectar; and, fortunately, we have calculated that 27.78 kg of nectar is necessary to obtain 25 kg of honey. Remember, we are required to find the quantity of mango necessary to obtain 25 kg of honey. We, therefore, represent the unknown quantity of mango by x and similarly proceed as above:

$$100 \text{ kg of mango} = 65 \text{ kg of nectar,}$$
$$x \text{ (kg) of mango} = 27.78 \text{ kg of nectar.}$$

In other words, 100 kg of mango is to 65 kg of nectar as x (kg) of mango is to 27.78 kg of nectar:

$$\frac{100}{65} = \frac{x}{27.78}$$

Cross multiplying, we have the following:

$$65x = 27.78 \cdot 100 \Rightarrow x = 2778 \div 65$$

$x = 42.74$ (kg) – the quantity of mango necessary to obtain 25 kg of honey

Second Method:
    (a) Nectar contains 90 % of honey. Let x to be the quantity of nectar required to obtain 25 kg of honey. Therefore, we say 90 % of x = 25 kg of honey.
    (b) We then solve the equation:

$$90 \text{ \% } x = 25 \Rightarrow 0.9x = 25$$
$x = 25 \div 0.9 = 27.78$ (kg) – the quantity of nectar necessary to obtain 25 kg of honey.

    (c) Mango contains 65 % of nectar. Let x to be the quantity of mango required to obtain 25 kg of honey (which is also equivalent to 27.78 kg of nectar as determined above). Therefore, we set up the equation:

$$65\% \text{ of } x = 27.78 \text{ kg of nectar.}$$

    (d) We solve the equation:

$$65 \text{ \% of } x = 27.78 \Rightarrow 0.65x = 27.78 \Rightarrow x = 27.78 \div 0.65$$
$x = 42.74$ (kg) – the quantity of mango necessary to obtain 25 kg of honey

1.1.4   Finding the Ratio of One Number to Another

Problem 1.   375 kg of sugar can be processed or obtained from 500 kg of sugar cane. What is the percentage of sugar by composition in the total mass of sugar cane?

Solution.   Let us look at some methods of solving this problem.

First method:
We need to find what part of the number 500 (the total mass of sugar cane) does the number 375 (the mass of sugar) constitute; in other words, we need to find the ratio of the mass of sugar to the total mass of sugar cane:
$$375 : 500 = \frac{375}{500} = \frac{75}{100} = 0.75.$$
We convert 0.75 into percents: $0.75 = 75\%$.
Answer: the percentage of sugar by composition in the total mass of sugar cane is 75%.

Second method:
 Since 500 kg is equal to 100% of the total mass of sugar cane,
$$\text{then 1\% is equal to: } \frac{500}{100} = 5 \text{ (kg)}.$$
If 5 kg makes up 1%, then 375 kg would make up: $\frac{375}{5} \cdot 1\% = 75\%$ (the answer).

Third method:
 Let 375 kg of sugar be equal to x %. Since 500 kg corresponds to 100%, and 375 kg corresponds to x %, then 1% will be equal to either $\frac{500}{100}$ or $\frac{375}{x}$.
From here, the solution of the problem is reduced to the following expressions:

$$\text{500 kg is to 100\% as 375 kg is to } x \%.$$

 In other words, the ratio of 500 to 100 is directly proportional to the ratio of 375 to x. This means that the expressions $\frac{500}{100}$ and $\frac{375}{x}$ are equal. Therefore, we can write and solve the following equation:

$$\frac{500}{100} = \frac{375}{x} \quad \Leftrightarrow \quad \frac{x}{375} = \frac{1}{5};$$
$$x = \frac{1}{5} \cdot 375 = 75 \text{ (\%)}.$$

It is left with the pupil to judge and apply the method of solution to the problem, which she or he considers more convenient.

Problem 2. Out of 80 students in the 9th grade class of Palm Groove Community Junior High School, 8 did not receive the passing GPA required for graduation. What percent of the 9th grade students received passing GPA for graduation?

Solution. We can use two methods to solve this problem.

First Method

From the condition of the problem, we know that 72 students did receive the passing GPA required for graduation. We determine what percent of the 80 students is one student: $\dfrac{100}{80} = 1.25$ %. In order to determine what percent of the 80 students is 72 students, it is necessary to multiply: $1.25 \cdot 72 = 90$ (%). Therefore, the answer to the problem is 90 %.

Second Method

Since according to the condition of the problem, 72 students received the passing GPA required for graduation. We can reason otherwise: 72 is what percent of 80? This is the conventional method used by many people in solving such problem.

$$72 = \frac{x}{100} \cdot 80 \ \Rightarrow x = \frac{72}{80} \cdot 100 = 90 \ (\%).$$

In applying the second method, we have simply found the ratio of 72 to 80; we then multiplied the ratio by 100 to convert it to percent, which is required by the problem.

Problem 3. A market woman produced 900 kg of farina from 1,200 kg of raw cassava. What percent is the shortfall (or decrease) in the weight of the produced farina over that of the raw cassava used?

Solution. The mass of the shortfall is: $1,200 - 900 = 300$ (kg). Therefore, we can now determine the percent of the shortfall:

$$\frac{300}{1200} = \frac{1}{4} = 0.25 = 25\% \ \text{(Answer)}.$$

What other method can you think of to solve this problem?

Problem 4. Tugbeh's mother bought for him 75 Christmas balloons of which are the following colors:
red – 15; green – 18; yellow – 3; black – 36; white – 3. Bledee's mother also bought for him 70 Christmas balloons of which are the following colors: red - 21; green – 14; yellow – 28; black – 7; white – 0. With the help of a table and diagrams, compare and determine the percent of each color of balloons each of the boys has.

Solution. We look at 3 methods of solution to this problem, using tabular and graphical representations of the data given in the problem and corresponding determination of percentages as required.

First method

In Table 1.1 below, we determine and compare the percent of each color of balloons for each boy. Tugbeh has 36 black balloons, 18 green balloons, and 15 red ones, which make up 48%, 24%, and 20%, respectively. He has the same number (3) of yellow and white balloons, each making up 4%. So we are sure that the total number of balloons he has (75) is equal to 100%. Bledee has 28 yellow balloons, 21 red balloons, 14 green balloons, and 7 black ones, which make up 40%, 30%, 20%, and 10%, respectively. Bledee does not have any white balloons. His total number of balloons (70) is equal to 100%. The percentage values were determined by taking the ratio of the number of each color of balloons to the total number of balloons of each boy and multiplying it by 100%. For example, the percent of Tugbeh's 36 black balloons is: $\frac{36}{75} \cdot 100\%$
$= 0.48 \cdot 100\% = 48\%$.

It is obvious that it seems somewhat difficult at a glance to understand and compare the data represented in a table. Sometimes a bar chart (or *bar graph*) may be used to overcome this difficulty.

Table 1.1

Na-mes	Total Number of balloons	%	Red Number	%	Green Number	%	Yellow Number	%	Black Number	%	*White Number*	*%*
Tugbeh	75	100	15	20	18	24	3	4	36	48	3	*4*
Bledee	70	100	21	30	14	20	28	40	7	10	0	0

According to the data given in the problem and those determined in Table 1.1, we can draw and make use of a bar graph in Fig. 1.1 to graphically represent and compare the percents of each color of balloons for each boy. The bar graph makes it much simpler and easier to see and compare the percentages of each color of balloons the boys have.

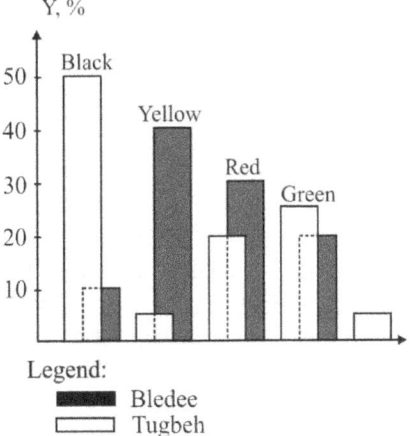

Legend:
  ■■■ Bledee
  ☐ Tugbeh

Fig. 1.1. Graphical representations to compare percentages of each color of balloons for each boy with the use of a bar graph.

Third method.

Besides the bar graph, we can draw a pie chart or pie diagram (also called *circular graph*) of the data represented in Table 1.1. Example of such a diagram or graph is given in Fig.1.2, which shows the distribution of percentages of each color of balloons for Bledee. (As a class work, each pupil is required to draw a circular graph of the distribution of percentages of each color of balloons also for Tugbeh, following the example in Fig.1.2.) As a circular graph, we draw a circle of an arbitrary radius which, as we know, contains 360^0. Thus 1% will make up 3.6^0; accordingly: $30\% = 108^0$; $20\% = 72^0$; $40\% = 144^0$; and $10\% = 36^0$. With the help of a protractor, we draw a circular graph in Fig.1.2 depicting the percentage distribution of the colors of Bledee's balloons

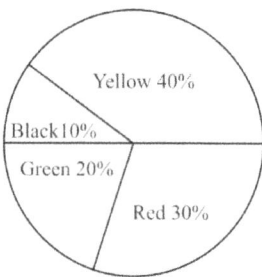

Fig.1.2. Distribution of percentages of each color of balloons for Bledee (one of the boys) with the use of a circular graph.

You will continue to meet bar graphs, circular graphs, and other similar diagrams not only in your study of mathematics, but also in other subject areas if and when the need arises to make comparisons between or among certain quantities. It is obvious that percentage facilitates comparison of different quantities. It is widely applied in sociology and demography, science and technology, statistics and economics, and other areas of academic and professional activities.

1.1.5 Simple and Compound Interests

Simple Interest

Sometimes, as circumstances dictate, it becomes necessary to borrow some money; that is, to obtain or receive some money on loan for temporary use, with the intention to give it, or something equivalent or identical, back to the lender. We all know that it is a common phenomenon in the business or commercial community to take a credit from a lender, usually a bank, for one reason or another. Usually, the borrower is required to pay to the lender some percentage as interest on the credit within a specified length of time. Also, money deposited into a bank by a person may accrue some interest over time. Before we proceed further, let us take a look at some basic definitions.

Interest is money paid as a percentage (which is called the rate) on a credit for a certain time period. In other words, *interest is a charge for the use of credit (borrowed money), in which such a charge is expressed as a percentage per time unit of the money borrowed or used.* The credit, or borrowed money, is usually referred to as the *principal. Principal is the capital borrowed on which an interest is paid; or it is the original sum of borrowed money on which interest is calculated.* The interest accrued on a principal may be considered as

22

either *simple* or *compound. Simple interest is interest calculated or paid on the principal alone. Compound interest* may be defined as *interest that is calculated on both the principal and its accrued interest.* The sum of the principal and the interest accrued on it is called the *amount.*

The basic formula for the calculation of interest is:

$$I = PRT$$

where I is the interest, P is principal, R is the rate expressed in percent, and T is the time period over which the interest accrues.

The principal may be calculated by the formula: $P = \dfrac{I}{RT}$; the rate may be calculated by the formula: $R = \dfrac{I}{PT}$; and the time as $T = \dfrac{I}{PR}$, when the interest, principal, and rate are known.

Problem 1. At the rate of 45 percent simple interest per annum, how much money would a bank deposit of $17,000.00 turn into within (1) 6 months? (2) 10 months? (3) 1 year?
Solution.

1. 6 months:
   (a) $I = PRT = 17,000 \cdot \dfrac{45}{100} \cdot \dfrac{6}{12} = 17,000 \cdot 0.45 \cdot 0.5 = \$3,825$ – the interest on $17,000 for 6 months at the rate of 45 % per annum;
   (b) $A = 17,000 + 3,825 = \$20,825$ – the amount generated for 6 months;

2. 10 months:
   (a) $I = PRT = 17,000 \cdot \dfrac{45}{100} \cdot \dfrac{10}{12} = 17,000 \cdot 0.45 \cdot \dfrac{5}{6} = \$6,375$ - the interest on $17,000 for 10 months at the rate of 45 % per annum;
   (b) $A = 17,000 + 6,375 = \$23,375$ - the amount generated for 10 months;

3. 1 year = 12 months:
   (a) $I = PRT = 17,000 \cdot \dfrac{45}{100} \cdot 1 = 17,000 \cdot 0.45 \cdot 1 = \$7,650.00$ - the interest on $17,000 for 1 year at the rate of 45 % per annum;
   (b) $A = 17,000 + 7,650 = \$24,650$ – the amount generated for 1 year;

Therefore, a bank deposit of $17,000 at 45 % per annum within 6 months will turn into $20,825, within 10 months - into $23,375, and within one year – into $24,650.

Problem 2. A certain bank pays out 20 percent simple interest annually on the deposits of its customers. How much will this bank pay to a customer who has deposited $5,000 for (a) 8 months? (b) two years?

Solution. To be provided independently by pupils.

Compound Interest

Compound interest is the concept of adding accumulated interest back to the principal, so that interest is earned on interest. The act of declaring interest to be principal is called *compounding*. In other words, interest is *compounded.* A loan, for instance, may have its interest compounded usually every three months (or a quarter of a year). That is, the compounding frequency is three months, or a quarter of a year. Thus a loan with $100 principal and 2% interest per quarter would yield a balance of $102 at the end of each quarter. In compound interest, the compounding frequency may also be annual (yearly), biannual (half-yearly), monthly, or daily.

Interest rates must be comparable in order to be useful. In order to be comparable, the interest rate and the compounding frequency must be disclosed. Most people regard interest rates as an annual percentage; therefore, many governments require financial institutions and individual lenders to disclose a *nominally* comparable annual interest rate on deposits, and/or borrowed money. For this reason, compound interest rates may be referred to as *annual percentage rate* or *effective interest rate.* Sometimes a fee is charged up front by an individual lender or financial institution to obtain a credit or loan. When such is the case, some governments usually require that the *annual percentage rate* counts on the cost of such fee as well as the compound interest in converting to the actual *equivalent* interest rate. These government requirements assist borrowers and consumers in general to more easily determine and compare the actual cost of borrowing money. In the same token, compound interest rates may be converted, in compliance with government regulation, to allow for comparison. Comparison is necessary because for any given compound interest rate and compounding frequency, there exists an **actual equivalent rate** of interest.

Although certain financial or business transactions may be subject to the calculation of simple interest, compound interest most often than not predominates in finance and economics. As we have previously noted, compound interest is contrasted with simple interest, where there is no compounding. The effect of compounding depends on the frequency with which interest is compounded and the effective periodic interest rate which is applied. Therefore, in order to accurately determine the actual amount to be paid with interest under legal terms, the compounding frequency (annual, biannual, quarterly, monthly, or daily) and the interest rate must be specified. Different conventions may be used from country to country, but in finance and economics, the principle is the same; and the usage of the following terms is common:

Periodic rate of interest is defined as the nominal annual interest rate divided by the number of compounding periods per year. The *periodic rate* (for which interest is charged and subsequently compounded) is used primarily for calculations, and is rarely used for comparisons.

Nominal interest rate or *nominal annual interest rate* is the annual rate, unadjusted for compounding. For instance, 8% nominal annual interest compounded on a quarterly basis has a periodic *quarterly* rate of 2%.

Effective annual rate is the annual rate *adjusted* to allow for comparisons. The nominal rate must be restated to reflect the effective rate as if annual compounding were applied.

Economists and financial experts generally prefer to use effective annual rates to allow for comparability. In finance, commerce or business, however, the nominal annual rate may be the most frequently used. When quoted with the compounding frequency, a loan with a given nominal annual rate is fully specified. In this way, the effect of interest for a given loan scenario can be precisely determined, but cannot be compared to loans with different compounding frequencies.

It is worthy of note that, despite the commonality of the aforementioned terms, their use may be inconsistent. That is, their use may vary depending on local customs, demand and supply realities, and other aggregates of marketing forces.

It is necessary to give an instance of how a bank interest rate works relative to the concept of *compounding*. As we have stated earlier, the *effective interest rate* depends on the number of compounding periods. For this reason, we look at an analysis of different scenarios of *compounding periods* given below:

(a) $r = 3\%$, yearly compounding: $100 invested for 1 year would yield $103; the *effective interest rate* is equal to 3%.

(b) $r = 3\%$, quarterly compounding: That is, the quarterly interest rate is $\dfrac{3\%}{4} =$ 0.75%. $100 invested for 1 year would yield $100 · $(1 + 0.0075)^4 = \$100 \cdot$ $(1.0075)^4 = \$103.03$; the ***effective interest rate*** is 3.03%.

(c) $r = 3\%$, monthly compounding: That is, the monthly interest rate is $\dfrac{3\%}{12} =$ 0.25%. $100 invested for 1 year would yield $100 · $(1 + 0.0025)^{12} = \$100 \cdot$ $(1.0025)^{12} = \$103.04$; the ***effective interest rate*** is 3.04%.

(d) $r = 3\%$, daily compounding: That is, the daily interest rate is $\dfrac{3\%}{365} =$ 0.008219178%; $100 invested for 1 year would yield approximately $100 · $(1 + 0.00008219)^{365} \approx \$100 \cdot (1.00008219)^{365} \approx \$103.05$; the ***effective interest rate*** is approximately equal to 3.05%.

In general, the ***effective interest rate*** may be calculated by the formula:

$$r_{eff} = (1 + \tfrac{r}{n})^n,$$

where n is the number of *compounding periods*, and r is the interest rate.

The mathematics of compound interest is more complex than that of simple interest, and may be beyond the scope of mathematics text intended for sixth grade pupils. Notwithstanding, the following material is produced for the purpose of acquainting pupils with the concept of calculating compound interest. The most basic formula is:

$$V_f = V_p \cdot (1 + r)^n,$$

where V_f and V_p are the *future value* and *present value*, respectively, of a sum; r is the interest rate expressed in percent; and n is the number of periods. The above formula calculates the future value V_f of an investment's present value V_p which accrues at a fixed interest rate of r for n periods.

From the above formula for V_f, we can derive the formulae for calculating V_p, r, and n. For example, the formula below

$$V_p = V_f \div (1 + r)^n$$

calculates what present value V_p would be needed to yield or produce a certain future value V_f if the interest rate r accrues for n periods. Also the formula

$$r = (V_f \div V_p)^{n-1} - 1$$

calculates the compound interest rate achieved if an initial investment of V_p returns a value of V_f after n accrual periods.

The formula for calculating n (the number of periods) required to get the future value V_f, when the present value V_p and the interest rate r are known, involves a logarithmic function which can be in any base (either log or ln):

$$n = (\log (V_f) - \log (V_p)) \div \log (1+r) .$$

The basic formula for calculating the amount of compound interest is:

$$A = P (1 + (r \div n))^{nt},$$

where P represents the principal amount (or initial investment); r represents the nominal annual interest rate expressed in % (or expressed as a decimal); n represents the number of times the interest is compounded per year; t represents the number of years; and A represents the amount after time t.

Problem 3. Mr. Kamara deposited an amount of $3,875.00 in a bank paying an annual interest rate of 6%, compounded biannually (half-yearly). Calculate the balance which the bank is supposed to pay to Mr. Kamara after 1 year.

<u>Solution:</u> Using the formula above, **P = $3,875.00; r = 6% = 0.06; n = 2; t = 1**, we calculate the amount **A**:

$A = P(1 + (r \div n))^{nt} = 3875(1 + (0.06 \div 2))^{2 \cdot 1} = 3875(1.03)^2 = $
$= 3875 \cdot 1.0609 = 4110.9875 \approx 4,110.99;$
$A = \$4,110.99$ – the balance which the bank is supposed to pay
to Mr. Kamara after 1 year.

Each time unpaid interest is compounded and added to the principal, the resulting principal is grossed up to an amount which is equal to $P(1 + r \%)$.

Problem 4. If you are told that the interest rate is 6% per annum, compounded biannually, calculate the *equivalent effective annual rate* of interest.

<u>Solution:</u> *Compounded biannually* means **n = 2**. The 6% is the nominal rate, which implies an effective biannual interest rate ($r = 6\% \div 2 = 3\% = 0.03$). Let us start with **P = $100**, for instance. Then at the end of one year (**t = 1**), the amount **A** would have accumulated to:

$A = P \cdot (1 + r \%)^{nt} =$
$= \$100 \cdot (1 + 0.03)^{2 \cdot 1} = \$100(1 + 0.03)^2 = \$106.09.$

We all know that $100 invested at the interest rate of 6.09% per annum will yield $106.09 at the end of one year. Therefore, *the equivalent effective annual rate* is **6.09%.** Therefore, the *equivalent effective annual rate* of interest is equal to 6.09 %.

1.2   Ratios and Proportions

1.2.1   Properties of Ratios

Problem 1. The population of Nigeria, for example, is about 160 million people, and that of Liberia is about 3 millions. How many times the population of Nigeria is greater than that of Liberia?

Problem 2. In a filling station, 1,000 liters of gasoline is required to fill a reservoir having a capacity of 1,200 m^3. What quantity of gasoline is needed to fill a similar reservoir having a capacity of 1,800 m^3?

The above problems are similar, and solutions to them require a similar approach related to the application of ratios and proportions.

In mathematics there is a convenient method of comparing quantities that are similar. In comparing such quantities, we try to find an answer to question like "how many times one quantity is greater than another quantity". The answer to such question is found with the help of division. The quotient of two numbers in such cases is called a *ratio*. A *ratio is a number indicating how many times one quantity is greater than another, or indicating what part of one quantity another quantity is made up.* The two numbers making up a ratio are called *members of the ratio*. The first member is said to be the *antecedent*, and the second member is called the *succeedent.* For example, in the ratio 9:8 (or $\frac{9}{8}$), the number 9 is the *antecedent* member, and the number 8 is the *succeedent* member of the ratio.

Let us look at the ratio 24:16. If we divide both members of the ratio by 4, then we will obtain 6:4. And if we multiply them by 5, we will have 120:80. By dividing both members of the original ratio by 4, or multiplying them by 5, does not change the value of the original ratio. All three ratios can be replaced by one ratio: 3:2. In a general form, it I possible to write this property of ratios as the following:

$$m : n = \frac{m}{n} = \frac{m \div k}{n \div k} = \frac{m \cdot k}{n \cdot k}$$

Let us look at some important properties of ratios:

1. A ratio of physical quantities can be replaced by a ratio of the numbers measuring these quantities, for example:
   (a) $33\text{m}^2 : 11\text{m}^2 = 33:11$;
   (b) $9\text{kg}:3\text{mg} = 9,000\text{g}:3\text{mg} = 9,000,000\text{mg}:3\text{mg} = 9,000,000:3$.

2. A ratio of larger numbers can be replaced by a ratio of small numbers. For example:
   (a) $500,000:25 = 20,000:1$;
   (b) $150:3 = 50:1$.

3. A ratio of fractional numbers can be replaced by a ratio of whole numbers, for example:
   (a) $8:\dfrac{1}{3} = 24:1$;
   (b) $7.4:4.5 = 7\dfrac{2}{5} : 4\dfrac{1}{2} = \dfrac{37}{5} : \dfrac{9}{2} = 2(37) : 5(9) = 74 : 45$.

1.2.2   Properties of Proportions

Proportions make use of ratios that are equal. In other words, *the equality of two ratios* is known as *proportion*. For examples:

(a) $3:4 = 6:8$;
(b) $6:1 = 24:4$;
(c) $50:100 = 2:4$;
(d) $\dfrac{4}{5}:\dfrac{2}{3} = \dfrac{4}{10}:\dfrac{2}{6}$;
(e) $0.25:3 = 0.5:6$;
(f) $3:2 = 12:8$;
(g) $320,000:64 = 10,000:2$.

With the help of letters, proportion can be expressed as:

$$a: b = m:n, \ or \ \frac{a}{b} = \frac{m}{n}$$

This expression may be read as: *"a divided by b is equal to m divided by n"*, or *"the ratio of a to b is equal to the ratio of m to n"*. In the proportion a:b =

m:n, letters ***a*** and ***n*** are called the extreme members; and letters ***b*** and ***m*** are known as the mean members of the proportion.

The proportion 1.4:5.6 = 7:28 is *true* because the values of its left and right parts are the same and equal to $\frac{1}{4}$. Let us find the product of the *extreme members* as well as the product of the *mean members* of this proportion:

$$a \cdot n = 1.4 \cdot 28 = 39.2; \qquad b \cdot m = 5.6 \cdot 7 = 39.2.$$

"If a proportion is true, then the product of its extreme members is equal to the product of its mean members". This property is known as the *basic property of proportion*.

<u>Proof</u>

Let us assume that the proportion *a:b* = *m:n* is *true*. If the value of its left part is equal to *p*, then the value of its right part is likewise equal to *p*:

$$a{:}b = p; \ and \ m{:}n = p.$$

It therefore follows from here that
$$a = bp \ \text{and} \ m = np.$$

We examine and compare the products of the *extreme* and *mean members* of the proportion:

$$an = bp \cdot n = bnp;$$
$$bm = b \cdot np = bnp.$$

It is easy to see that the product of the extreme members of the proportion is equal to the product of its mean members: if *a:b* = *m:n*, then *an* = *bm*.

With the help of proportion, it is possible to solve many problems.

1.2.3   Direct and Indirect Proportionalities

Problem 1. A market woman sells 3 oranges for 7 ¢.
       (a) What would be the cost of 21 oranges?
       (b) What would be the cost of 1 orange?
       (c) What would be the cost of *n* oranges?
Solution.

(a) This is an ordinary proportion problem. The answer can be found by applying the basic property of proportion. If we let the unknown (the cost of 21 oranges) to be x, then we can set up the solution to the problem like this: 3 is to 7 as 21 is to x; that is,

$$3:7 = 21:x, \text{ or } \frac{3}{7} = \frac{21}{x} \Rightarrow 3 \cdot x = 7 \cdot 21;$$
$$3x = 147 \Rightarrow x = 147 \div 3 = 49;$$
21 oranges will cost 49 ¢ (answer).

(b)

$$3:7 = 1:x, \text{ or } \frac{3}{7} = \frac{1}{x} \Rightarrow 3 \cdot x = 7 \cdot 1;$$
$$3x = 7 \Rightarrow x = 7 \div 3 = \frac{7}{3} = 2\frac{1}{3};$$

1 orange will cost $2\frac{1}{3}$ ¢ (answer).

(c) The same method can be applied here. If we let the unknown (the cost of n oranges) to be x, then we can likewise set up the solution: 3 is to 7 as n is to x;

$$3 : 7 = n : x, \text{ or } \frac{3}{7} = \frac{n}{x} \Rightarrow 3 \cdot x = 7 \cdot n;$$
$$3x = 7n \Rightarrow x = 7n \div 3;$$
$$x = 2\frac{1}{3}n \ (¢);$$

n oranges will cost $2\frac{1}{3}n$ ¢ (answer).

Problem 2. With a constant speed, a motorist drives 82 miles for two hours. Calculate his speed in *mph*. At this speed, calculate the respective distances he can cover within 3, 4.5, 6, and 7 hours and register the results in Table 1.2.
Solution. We first calculate the constant speed of the motorist in *mph*:

$$v = \frac{d}{t} = \frac{82}{2} = 41 \text{ (mph)}.$$

Table 1.2

Speed v, mph	41	41	41	41	41	41
Time t, Hours	1	2	3	4.5	6	7
Distance d, miles	41	82	123	184.5	246	287

In registering the results of calculation into Table 1.2, it is easy to observe that the *distance (d)* of motion is equal to the product of the *speed (v)* and the *time (t)*, that is $d = v \cdot t$. Also observe that, as Table 6.2 indicates, *distance (d)* and *time (t)* are variable quantities; but their ratio ($\frac{d}{t}$), the speed of motion, is a constant in the given case and equal to *41 mph*:

$$v = \frac{d}{t} = \frac{41}{1} = \frac{82}{2} = \frac{123}{3} = \frac{184.5}{4.5} = \frac{246}{6} = \frac{287}{7} = 41.$$

If two quantities changes such that the ratio of the corresponding values of their quantities is a constant, then such quantities are said to be directly proportional. In other words, *if two variable quantities are connected with each other such that with an increase (or decrease) in the value of one of them by a number of times, the value of the other quantity is increased (or decreased) by as much times, then such quantities are said to be in direct proportionality.*

In the solution to Problem 2 above, as indicated in Table 1.2, *distance (d) and time (t)* are directly proportional. What other variable quantities do you know that are directly proportional?

Problem 3. With what speed must a motorist drive so as to travel 60 miles in 1 hour? Calculate the respective speeds with which he must drive in order to cover the given distance within 2, 3, 4, 5, and 6 hours and register the results in Table 1.3.

Solution. We calculate the speed of the motorist in 1 hour:
$$v = \frac{d}{t} = \frac{60}{1} = 60 \ (mph).$$

Table 1.3

Distance d, miles	60	60	60	60	60	60
Time t, Hours	1	2	3	4	5	6
Speed v, mph	60	30	20	15	12	10

It can be seen from Table 1.3 that the *speed (v) and time (t)* are variable quantities; but the product $(v \cdot t)$ of their corresponding values is a constant and equal to 41:

$$v \cdot t = 60 \cdot 1 = 30 \cdot 2 = 20 \cdot 3 = 15 \cdot 4 = 12 \cdot 5 = 10 \cdot 6 = 60.$$

If two quantities changes such that the product of the corresponding values of their quantities is a constant, then such quantities are said to be indirectly or inversely proportional. In other words, *if two variable quantities are connected with each other such that in increasing (or decreasing) the value of one of them by some times, the value of the other quantity is decreased (or increased) by as much times, then such quantities are said to be in indirect or inverse proportionality.*

In the solution to problem 3 above, as indicated in Table 1.3, *speed (v)* and *time (t)* are indirectly or inversely proportional. What other variable quantities can you think of that are indirectly or inversely proportional?

Chapter One: Questions to Test Your Understanding

1. Define the following terms:
   (a) Percent;
   (b) Ratio;
   (c) Antecedent and succeedent;
   (d) Proportion;
   (e) Direct Proportionality;
   (f) Inverse Proportionality;
   (g) Extreme members of a proportion;
   (h) Mean members of a proportion.

2. Answer the following questions:
   (a) How can we convert a common fraction into percent?
   (b) How can we convert a decimal fraction into percent?
   (c) How can we convert a percent into a common fraction?
   (d) How can we convert a percent into a decimal fraction?
   (e) Give an example of questions (a), (b), (c), and (d).

3. How can we find a percent of a given number? Give an example.

4. How can we find a number by its given percent? Give an example.

5. Explain in your own words:

   (c) What is a bar graph, and how can it be drawn?
   (d) What is a circular graph, and how can it be drawn?

6. State three properties of ratios that you know.

7. What is said to be the basic property of proportions?

8. How can we find an unknown *mean* member of a proportion?

9. How can we find an unknown *extreme* member of a proportion?

10. From which ratios is it possible to make up a proportion?

11. Define the following terms:
  (a) Interest
  (b) Rate of interest
  (c) Principal
  (d) Simple interest
  (e) Compound interest
  (f) Amount

12. What is the basic formula for the calculation of interest?

Chapter One: Problems and Exercises

1. Express the following percents in the form of common and decimal fractions.

   Example: $50\% = \frac{1}{2} = 0.5$.

   (a) (i) 9 %; (ii) 500 %; (iii) 5.5 %; (iv) 508.8 %; (v) 175 %.
   (b) (i) 25 %; (ii) 270 %; (iii) 15.25 %; (iv) 3.3 %;
   (c) (i) 87 %; (ii) 200 %; (iii) 25.75 %; (iv) 200.1 %;
   (d) (i) 100 %; (ii) 108 %; (iii) 9.2 %; (iv) 0.05 %; (v) 135 %.

2. Express each of the following in percent:

   (a) (i) $\frac{9}{100}$; (ii) $\frac{37}{100}$; (iii) $\frac{27}{25}$; (iv) $\frac{9}{500}$;

   (b) (i) $\frac{3}{5}$; (ii) $\frac{1}{4}$; (iii) $\frac{16}{600}$;

   (c) (i) $\frac{3}{25}$; (ii) $\frac{7}{50}$; (iii) $\frac{9}{10}$;

   (d) (i) $\frac{3}{50}$; (ii) $\frac{9}{20}$; (iii) $\frac{475}{300}$;

   (e) (i) $\frac{5}{4}$; (ii) 3.91; (iii) 5.2065

3. Find the following:
   (a) 15 % of $150;
   (b) 25 % of 600 apples;
   (c) 300 % of $300;
   (d) 0.5 % of 50.

4. Compare each pair of expressions, which is greater?
   (a) 44 % or 0.44;
   (b) 5.5 or 5.5 %;
   (c) $\frac{7}{10}$ or 70 %;
   (d) 8.5 or 85 %.

5. The Accra Bank for Investment & Development pays out 14 % per annum on deposits to its customers.
   (a) How much compound interest did the bank pay out to a depositor for 3 years, if the deposit was $6,500 United States dollars?
   (b) How much did the bank pay out in simple interest to a depositor for 9 months, if the deposit was $1,200 United States dollars?

6. Lemon grass loses 65 % of its weight when it is dried. What quantity of dry lemon grass can be obtained from 60 kg of fresh lemon grass?

7. Seven oranges can produce a quantity of juice equivalent to 200g, which is 40 % of a bottle of natural orange juice. How many oranges are required to produce six such bottles of natural orange juice? What is the mass of one such bottle of natural orange juice?

8. A passenger ship and a ferry boat depart simultaneously from two ports, traveling in opposite direction towards each other. The distance between the ports is 270 miles, and the speed of the boat is 30 mph; the speed of the ship is 80 % that of the boat. Within how many hours will the ship and boat meet?

9. After a technician manufactured 660 components on a production line, the number of components yet remains to be manufactured make up 45 %. Calculate how many components yet remain for him to manufacture?

10. Kamara's score on his Math exam is 75. His score in the Science exam is 110% of his score in Math. His score in History exam is 60 % of the sum of his scores in both Math and Science. Calculate the arithmetic mean of his scores in the three subjects.

11. Ali has 80 balloons 25 % of which are purple. Mohammed has 25 balloons 80 % of which are purple. Which of the boys has more purple balloons?

12. What will the result be if the number 12 is increased by 16 % and then the result decreased by 16 %?

13. Find a number if 5 % of it is equal to 5.

14. 60 % of virgin coconut oil (VCO) can be produced from coconuts. What quantity of coconuts is necessary in order to produce 180 kg of VCO?

15. In the preparation of fresh palm butter from palm nuts, 90 percent of the mass of palm nuts go as waste. What quantity of palm nuts is necessary in order to prepare 2.5 kg of fresh palm butter?

16. Fresh lemon grass contains 75 % of water, and dry lemon grass contains 10 % of water. What quantity of dry lemon grass can be obtained from 35 kg of fresh lemon grass?

17. In the 6th grade class, 55 % of the total number of pupils is girls. The number of boys in the class is 90. What is the ratio of girls to boys in the class?

18. A boy bought 3 items from a store. The price of the first item is 50 % of the total sum of money he paid for the 3 items. The price of the second item is 10 % 0f the price of the first. The price of the third item is $270. What is the total sum of money he paid for the 3 items?

19. Fresh boney (fish) contains 70 % of water and dried bony (fish) contains 10 % of water. What quantity of dried bony may be obtained from 60 kg of fresh bony?

20. What percent of 7 is equal to 7?

21. What percent of a number is equal to two times of the given number?

22. Solve the following problems:
    (a) What percent of one foot is 3 inches?
    (b) What percent of one meter is 8 cm?
    (c) How many times did a number increase, if it was increased by 500%?

(d) A number was increased by 75 %. By how many times was it did increase?

23. By how many percents a number increases, if it is increased by:
    (a) 4 times?
    (b) 1.3 times?
    (c) 1.75 times?
    (d) 1.5 times?

24. By how many times a number decreases, if it is decreased by:
    (a) 25 %?
    (b) 60 %?
    (c) 99 %?

25. By how many percents a number decreases, if it is decreased by:
    (a) 0.9 times?
    (b) 12 times?
    (c) 5 times?

26. Solve the following:
    (a)   84 is what percent of 420?
    (b) From a coffee bag 0f 51 pounds, 3 pounds were picked and discarded as rotten coffee seeds. What percent of the coffee bag was good?

27. Solve the following:
    (a) 215 is what percent of 9850?
    (b) 465 is what percent of 5?
    (c) One number is 35 % of a second number. The second number is what percent of the first?

28. What percent of the Ghanaian population lives in the capital city Accra? The population of Ghana is 23 millions and the population of Accra is 3 millions (2005 estimate).

29. By how many percent does change the area of a rectangle with measurements of 40 inches and 150 inches, if the larger side is decreased by 50 % and the smaller side increased by 25%?

30. Table 1.4 below represents the average scores or marks in their subjects of the 9th grade students of the Mountain View Junior High School. Fill in the table.

Table 1.4

Characteristics of the class	Average score				Total number of students
	96	81	73	60	
Number of students	10	25	20	5	60
By Percent					
Percent of students achieving progress (good results)					

31. Draw a bar graph of the distribution of land area of the Earth among the given geographical regions below, using the data in Table 1.5.

Table 1.5

Item	Geographical region	Land area, km^2	Population,
1	Asia	43,810,000	4,000,000,000
2	Africa	30,370,000	933,500,000
3	North America	24,490,000	515,000,000
4	South America	17,840,000	371,000,000
5	Antarctica	13,720,000	0
6	Europe	10,180,000	810,000,000
7	Australia	7,600,000	20,000,000
8	Oceania	1,322,000	16,000,000
Total		149,332,000	6,665,500,000

32.   Draw a circular graph (or pie diagram) of the distribution of the world's population among the given geographical regions, using the data in Table 1.5 above.

33.   Rice loses 30% of its mass during the process of milling it. What quantity of rice is necessary to mill so as to get 50 kg of rice after milling?

34.   The following crops were planted on an experimental plot of farm land: rice = 600 hectares; cassava = 200 hectares; eddoes = 500 hectares; yams = 400 hectares; pumpkins = 300 hectares. Represent this data on either a bar or circular graph, depending on which one is easier for you.

35.   A certain grade of iron ore in Liberia contains 73% of iron. What quantity of this ore is required to get 73 tons of iron?

36.   The basic ingredients used to prepare a banana cocktail are 2kg of banana, 4kg of powder milk, 3kg of water, 300g of sugar, and 200g of dry cream. Calculate the percentage composition of each ingredients of the cocktail.

37. Find the indicated ratios:

(a) 0.12 to 0.6;   (b) 84 to 7;   (c) $\frac{1}{3}$ to $\frac{1}{6}$;

(d) $1\frac{2}{5}$ to $1\frac{1}{4}$;   (e) 5cm : 5mm;   (f) 20kg : 200g;

(g) 25m : 500cm;   (h) 15min : 1 hr.

38. Find the unknown member in each of the following expressions:

(a) $5 : x = \frac{1}{2}$;   (b) $x : 3 = \frac{1}{4}$;   (c) $x : 0.25 = \frac{3}{4}$;

(d) $0.5 : x = \frac{3}{8}$;   (e) $9 : x = \frac{1}{3}$;   (f) $\frac{250}{x} = 100$;

(g) $\frac{x}{10} = 1.5$;   (h) $\frac{8.1}{x} = 0.9$

39. A light pole having the height of 15.25m casts a shadow with Length 25.5m. Find the ratio of the pole's height to the length of its shadow.

40. How does a ratio change, if its antecedent member is decreased: (a) 4 times? (b) 15 times?

41. How does a ratio change, if its antecedent member is increased 6 times and succeedent member increased 9 times?

42. How does a ratio change, if its antecedent member is decreased 12 times and succeedent member decreased 48 times?

43. How does a ratio change, if the antecedent and succeedent members are: (a) divided by 2?   (b) multiplied by 9?

44. Replace each ratio of fractional numbers by a ratio of whole numbers:

(a) $\dfrac{3}{4} : \dfrac{1}{2}$;

(b) $\dfrac{1}{4} : \dfrac{4}{5}$;

(c) $2\dfrac{1}{2} : 3\dfrac{1}{4}$;

(d) $0.6 : 0.05$;

(e) $\dfrac{3}{4} : 0.24$;

(f) $\dfrac{4}{7} : \dfrac{6}{7}$.

45. On a map each 250, 000m is represented by 50mm. Determine the numerical scale to which the said map is drawn.

46. What is the length of a 75-km distance on a map, if the numerical scale of the drawing is $\dfrac{1}{1,000,000}$?

47. The length and width of a garden fence are 105 miles and 70 miles, respectively. What will its dimensions (length, width, and area) be on a plan drawn to $\dfrac{1}{350,000}$ scale?

48. On the floor plan of a building drawn to $\dfrac{1}{32}$ scale, the area of a hall is $\dfrac{3}{4}$ ft². What is the real (natural) area of this hall?

49. Determine whether or not it is it possible to make up a proportion from such ratios?

(a) 17:51 and 7:21;

(b) $3\dfrac{1}{8} : 2\dfrac{1}{4}$ and $2\dfrac{1}{4} : 3\dfrac{1}{8}$.

50. Find the unknown member in each of the following proportions:

(a) $3 : 5 = x : 6$;

(b) $x : 35 = 70 : 350$;

(c) $2\dfrac{3}{5} : x = 1\dfrac{1}{2} : 3\dfrac{4}{5}$;

(d) $1.2 : 8.4 = x : \dfrac{3}{4}$;

51. Twelve gallons of palm oil were produced from 225 kg of palm nuts. How many gallons of palm oil may be produced from 10 kg of palm nuts?

52. For the sewing of 5 suits in a tailor shop, 25 yards of cloth were used. How many of such suits can be sewn from 85 yards of such cloth?

53. Determine the value of x in each of the following proportions:

(a) $9.5x : 4 = 6 : 11$;

(b) $1\dfrac{3}{4}x : \dfrac{7}{8} = 4\dfrac{1}{4} : \dfrac{1}{16}$;

(c) $5x : 12 = 3\frac{1}{2} : 5\frac{1}{4}$;   (d) $81 : 3x = 27 : 9$.

54. A secretary typed 250-page document in 6 hours 30 minutes. In what time can she type 750 pages of similar document?

55. For the feeding of a kindergarten school the school administration procured food for 250 days at the outlay rate of 750 kg of food per day. For how many days the food could suffice or last, if the daily feeding (outlay of food) of the school amounted to 500 kg?

56. 25 railway coaches each with 20-ton loading capacity are required for transportation of iron ores from Yekepa to Monrovia. How many 15-ton railway coaches would be required for the transportation of said quantity of iron ores?

57. It was decided to replace the old rails (each 5 miles in length) on a section of the Monrovia-Yekepa rail way by new ones (each 7 miles in length). How many new rails would be necessary in order to replace 350 old rails?

58. The ingredients used in baking a certain kind of bread are corn meal, coconut oil, eggs, sugar, and cream in the ratio of $5 : 4 : 3 : 2 : 1$. How much of each ingredient by mass is necessary to take in order to bake 75 kg of this kind of bread?

59. A certain alloy is composed of iron, aluminum, and copper, taken respectively in the ratio of $3 : 1 : 2$. How much of each of these substances by mass is required so as to obtain 900 kg of this alloy?

60. By how many times the area of a square would increase if its side is increased 4 times? Illustrate your answer by a drawing of an arbitrary square.

61. By how many times the area of a rectangle would increase, if one of its sides is increased 3 times?

62. By how many times the volume of a cube would increase, if its edge is increased 5 times?

63. A motorist travels the distance between Abidjan and Accra in 10 hours with an average speed of 25 mph. Within which time interval the motorist can travel this distance, if his speed were increased by 15 mph?

64. A 25-lb load is suspended from one arm of a lever, length of which is 45 inches. Find the value of a weight that could be suspended from the other arm of the lever (having a length of 15 inches) to balance this load.

65. A group of 5 boys mowed 1,500 m^2 of lawn for 9 working days. What area of lawn a group of 3 boys could mow under identical conditions for 24 working days?

66. 24.9 tons of food is required to feed 83 children in a nursery for 60 days. What quantity of food would be required to feed 250 children in the nursery for 100 days under the same conditions?

67. On the first day a secretary typed 25% of the total number of pages of a document she was instructed by her boss to do. On the second day she typed $\frac{3}{8}$ of the total number of pages of the document; and on the third day she typed the remaining 75 pages of the document.
    (a) What percent of the document did she type each day?
    (b) What is the total number of pages in the document?
    (c) How many pages of the document did she type each day?

Chapter Two:
More about Geometric Concepts

2.1   Definitions of Geometric Terms

Geometry is the branch of mathematics which deals with the properties, relationships, and measurements of *points*, *lines*, *curves*, and *surfaces*. This chapter is intended as both a review and continuation of previous discussions concerning basic geometric concepts. In doing so, it attempts to build on those concepts to expand the discussion to *polyhedrons*, other kinds of *quadrilaterals*, and *circular bodies*, which are yet unfamiliar to pupils from the standpoint of materials already treated within the context of this book.

A *polyhedron* is a *solid figure* consisting of four or more plane (flat) *faces* (all polygons), pairs of which meet along an *edge*, with three or more edges meeting at a *vertex*. *Pyramids, cubes,* and *prisms* are examples of polyhedrons. In a *regular polyhedron* all the *faces* are identical regular polygons making equal angles with each other. In other words, a *face* of a polyhedron is bounded by a circuit of *edges*, and is usually a plane area which describes a *polygon*. The faces together constitute or make up the surface of the polyhedron. An *edge* (usually a *straight line* segment) joins one *vertex* to another and one face to another. The edges together make up the frame or *skeleton* of the polyhedron, in which a *vertex* forms a corner point. Specific polyhedrons (such as *tetrahedrons, pentahedrons, hexahedrons, heptahedrons, octahedron, nonahedron, decahedron, icosahedrons, icosatetrahedrons,* etc.) are named according to the number of their *faces*.

For example, a *tetrahedron* is a polyhedron having four plane faces. A *regular tetrahedron* has faces that are equilateral triangles. A *pentahedron* is a polyhedron which has five plane faces. A *hexahedron* is one in which there are six plane faces. A *heptahedron* is one in which there are seven plane faces; and so on. *Icosahedron* is a polyhedron having twenty faces. The faces of a *regular icosahedron* are also equilateral triangles. *Icosatetrahedron* is a polyhedron having twenty-four *trapezoid* faces, as occurring in some crystals.

We have also previously talked about *polygons*. A *polygon* is a closed plane figure bounded by three or more straight sides (or *straight line segments*) that meet in pairs in the same number of vertices, and do not intersect other than at these vertices. The sum of the degree measure of the interior *angles* of a *regular polygon* is $(n - 2) \cdot 180$ for n sides. The sum of the *exterior angles* of a regular polygon is *360 degrees*. A *regular polygon* is one that has all its *sides* and *angles* equal. Specific polygons are named according to the number of their sides, such as *triangle* (3 sides), *tetragon* (four sides), *pentagon* (five sides), *hexagon* (six sides), *heptagon* (seven sides), *octagon* (eight sides), *nonagon* (nine sides), *decagon* (ten sides), and so on.

Tetragon is a less common name for *quadrilateral*, which is a polygon having four sides. Examples of quadrilaterals are *rectangles* and *squares*. There are other types of quadrilaterals such as *parallelograms, trapezoids, etc.* A *parallelogram* is a quadrilateral whose opposite sides are parallel and equal in length. A *trapezoid* is a quadrilateral having one or neither pair of sides *parallel.*

In the study of geometry, we frequently make use of terms such as *points, lines and line segments, perpendicular and parallel lines, vertex* (plural *vertices* or *vertexes*), *angles, interior and exterior angles, vertically opposite angles, planes, diagonals,* etc. As the object of this section, we have defined below a few of these terms, some of which you already have been acquainted with.

A *point is a geometric category which describes a specific location within a given space that consists of neither area, length, volume, nor any other higher dimensional analogue.* In other words, a *point* is a geometric attribute having no dimensions and whose position in space is determined by means of its coordinates. In this way, a point is a *zero-dimensional* object. A *point* is basically one of the simplest, if not the simplest, geometric concepts – since it is often used in one form or another as the fundamental constituents of geometry, physics, vector graphics, etc.

A *line* may be formally defined as *any straight one-dimensional geometric element whose identity is determined by two points.* This suggests that exactly one line can be found that passes through any two points; and it provides the shortest connection between the points. A *line* may also be described as *an ideal zero-width, infinitely long, perfectly straight curve containing an infinite number of points.* The term *curve* in mathematics includes *straight curves.*

In two dimensions, *lines* in a *Cartesian plane* can be described algebraically by *linear equations* and *linear functions.* Thus, in this context, a *line* can be defined as a set of points (x, y) that satisfy the *characteristic equation*:

$$y = mx + c,$$

which is often given by the *slope-intercept form,* where m is the *gradient* or *slope* of the line, c is the *y-intercept* of the line (that is, the y-coordinate of the point where the line crosses the *y-axis*), and x is the *independent variable* of the function y. Naturally, the discussion of lines in a two-dimensional Cartesian plane geometry involving the use of linear equations and functions as well as the slope-intercept form is beyond the scope of this book, except for above-average pupils who out of self-motivation and intellectual curiosity may be interested in the topic. The full treatment of the topic also hinges on the country-specific requirements of a national sixth grade mathematics curriculum. However, it should be noted that as *two-dimensional* geometric elements, two different lines can either be *parallel*, or may intersect at *one and only one* point. In three or more dimensions, lines may also be *skew*, meaning *they don't meet, but also don't define a plane*; and that two distinct planes can intersect in *one and only one line.*

A *line segment* is a part of a line that is bounded by two distinct end points, designated by letters, and contains every point on the line between its end points. Depending on how the line segment is defined, either of the two end points may or may not be part of the line segment. Two or more line segments may have some of the same characteristics as lines, such as being *parallel, intersecting, or skew.*

Parallel lines refer to two or more lines which are separated by an equal distance at every point. In other words, parallel lines or sides never intersect or meet each other. Examples of parallel lines are the opposite edges of a ruler, textbook, rail lines, etc. What other parallel lines can you think of?

A vertex is a corner point of a polygon, polyhedron or other higher dimensional analogue, formed by the intersection of edges and faces of the object. Put simply, a *vertex* is the point of intersection of two sides of a plane figure or angle. A *vertex of a polygon* is the point of intersection of two edges; a *vertex of a polyhedron* is the point of intersection of three or more edges or faces.

A perpendicular (or perpendicular line or plane) is a line or plane which intersects another line or plane at right angles. In other words, two lines or two planes (or a line and a plane) are considered *perpendicular* to each other if they form right angles in their intersection. Conversely, any two lines that meet or intersect each other to form right angles are *perpendicular*. As we have discussed earlier, right angles measure ninety degrees (90^0).

Let us imagine, for example, that a horizontal line **KL** and a vertical line **DE** meet each other at point **E** to form right angles **DEL** and **DEK**. See Fig.2.1. Point **E** divides **KL** into two equal parts. Thus, referring to the drawing described below, we can say that line **DE** is *perpendicular* to line **KL** through the point **E**. It should be noted that, by definition, *a line is infinitely long*, and strictly speaking DE and **KL** in this example represent line segments of two infinitely long lines. Therefore, the line segment **DE** does not even have to touch or intersect line segment **KL** to be considered *perpendicular lines*. DE could be suspended over **KL**, without touching **KL**. If the line segment **DE** is extended out to infinity, it would still form adjacent right angles with line segment **KL**.

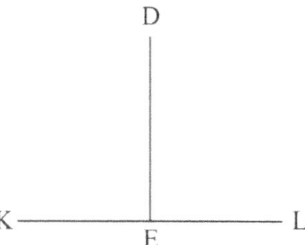

Fig. 2.1. An example of perpendicular line segments forming adjacent right angles.

We have been using the word *plane* without a formal definition. In mathematics, a *plane* may be considered as *a theoretical surface which has infinite width and length, zero thickness, and zero curvature.* A simple polygon divides a plane into two regions – the region inside it (the polygon) and the region outside it. Angles described or formed by the meeting (or connection) of the sides of a polygon within the polygon are called *interior (or internal) angles.*

An *angle* is a figure formed by two *rays* sharing a common *end point*, which is called the *vertex* of the angle. The magnitude of an angle is the *degree of rotation* that separates the two rays, and can be measured by considering the length of circular arc swept out when one ray is rotated about the vertex to coincide with the other ray. In a previous section, we have discussed the method by which an angle is measured. The term *angle* is used interchangeably for both the geometric figure itself and for its angular magnitude, which is a numerical quantity. The word *angle* is derived from the Latin word *angulus*, which means *a corner*. You also can remember that we earlier discussed several types of angles based on their measurement in degrees: *straight angle* (also referred to as an *open angle*), which is a straight line in essence; *obtuse angle*; *acute angle*; *right angle*; *supplementary* and *complementary angles.*

An *interior angle* is an angle of a polygon contained between two adjacent sides, lying inside the region within the polygon between the two intersecting sides. In other words, an interior angle is formed by two sides of a simple polygon sharing an endpoint. *A simple polygon has exactly one interior angle by vertex.* By contrast, an *exterior angle* is an angle of a polygon contained between one side extended and the adjacent side, and is outside the region between the two intersecting lines. It is an angle formed by one side of a simple polygon and a line extended from an adjacent side.

For example, the total measure of degrees of the *interior angles* in a five-sided regular polygon (or pentagon) is:

$$(n-2) \cdot 180^0 = (5-2) \cdot 180^0;$$

$$= 3 \cdot 180^0 = 540^0.$$

If we divide the total measure of degrees of the *interior angles* in a regular pentagon by the number of sides (that is, by 5) to find the degree measure of each angle, then *each interior angle in a regular pentagon is* 108^0. We can also easily use another method which is based on the measure of an exterior angle. Since it is known that every regular polygon can be built from *n isosceles triangles,* to get the measure of each *interior angle* we simply subtract the measure of an exterior angle from 180^0:

$$180 - (360 \div 5) = 180 - 72 = 108^0.$$

It is known that the sum of the *exterior angles* of a regular polygon is *360 degrees*. Consequently, the measure of degrees of each *exterior angle* in a regular pentagon is determined by dividing 360^0 by the number of sides the polygon contains (in this case, by 5):

$$360 \div 5 = 72^0.$$

Therefore, the degree measure of each *exterior angle* in a regular pentagon is 72^0.

A *diagonal* is a line connecting any two vertices that in a polygon are not adjacent and in a polyhedron are not in the same face.

Vertically opposite angles refer to the pair of equal angles between a pair of intersecting straight lines. Vertically opposite angles are equal.

2.2   Parallel and Perpendicular Lines

In a section of this book we talked about *parallel lines*. It should be noted that a basic characteristic of a line is that *it is considered limitlessly extendable in both directions*. Here and in any other section of this book, when we mention the word "line", we are referring to a "straight line", unless otherwise indicated.

In Fig.2.2, three lines (EF, GH, and IJ) are shown. From the figure it is easy to see that lines EF and GH do not intersect or meet each other and will never intersect or meet each other. Do you know why? It is because they lie in the same plane and are separated by an equal distance at every point. Accordingly, such lines are called *parallel lines.* The parallelism of two lines is symbolized by a pair of vertical lines (∥). Hence in the case of lines EF and GH, it is customary to write "EG ∥ GH" which reads "line EF is parallel to line GH", or "EF is a parallel of GH". Of course, if EF is parallel to GH, then GH is also parallel to EF.

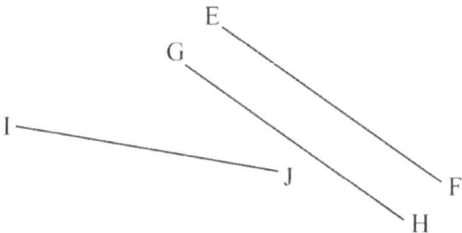

Fig. 2.2. Parallel and non-parallel lines.

On the other hand, it is obvious that lines IJ and GH would intersect or meet each other at one point if they are extended. This suggests that IJ and GH are *not separated by an equal distance at every point.* Consequently, line IJ is not parallel to line GH, or vice versa. Therefore, lines IJ and GH are said to be *non-parallel lines.*

You may be able to think about and name many objects around you (in your sleeping room, kitchen, class room, or in the street) which describe or exemplify parallel and non-parallel lines.

Rule 1: *If line **A** is parallel to line **B** and a third line **C** is not parallel to line **B**, then line **C** is said not to be parallel to line **A**.* Accordingly, any line that is parallel to line **C** is said not to be parallel to both lines **A** and **B**. Therefore, as shown in Fig.2.2, line **IJ** is likewise not parallel to line **EF**, or vice versa.

This statement is simple reasoning and does not require any proof at this level. However, in your study of geometry, you shall get acquainted with one type of reasoning or another; and you shall be required to use and, in certain cases, prove some reasoning as you advance in your study of mathematics and science.

The two common types of reasoning are *induction* and *deduction*. *Induction* is a process of reasoning by which a general conclusion is drawn from a premise or set of premises based on experience or experimental evidence. *Deduction* is a process of reasoning from a general to a specific or particular conclusion. As stated, *inductive reasoning* proceeds from a specific to a general conclusion, where as *deductive reasoning* proceeds from a general to a specific conclusion.

As we have said before, a pair of two non-parallel lines lying in a plane can be extended to intersect each other at a point. When this happens, the angles formed at the point of their intersection are said to be *vertically opposite*. *Vertically opposite angles* share a common vertex and are bounded by the same pair of lines; but they are opposite to each other. Such angles are equal in magnitude, and are said to be *congruent*. The word *congruent* means "similar, or having something in common". It should be noted that *vertically opposite angles* may also be simply referred to as either *vertical angles* or *opposite angles*. Examples of *vertically opposite angles* are represented below in Fig. 2.3.

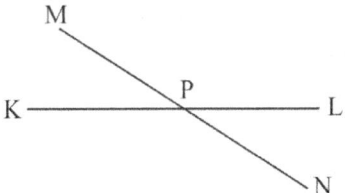

Fig. 2.3. Vertically opposite angles.

In Fig. 2.3, line MN intersects horizontal line KL at point P to form four angles: ∠MPL, ∠MPK, ∠KPN, and ∠NPL. These angles can be grouped into two pairs of *vertically opposite angles*: one pair consists of ∠MPL and ∠KPN (∠MPL = ∠KPN); and the second pair consists of ∠MPK and ∠NPL (∠MPK = ∠NPL).

Fig. 2.4 below shows four pairs of *vertically opposite angles*, in which line MN intersects two parallel lines PQ and RS. MN is said to be a *transversal line*. A *transversal line* (also simply referred to as *transversal*) is a line which passes through two or more other coplanar lines at different points. Name the four pairs of vertically opposite angles that are represented in Fig. 2.4.

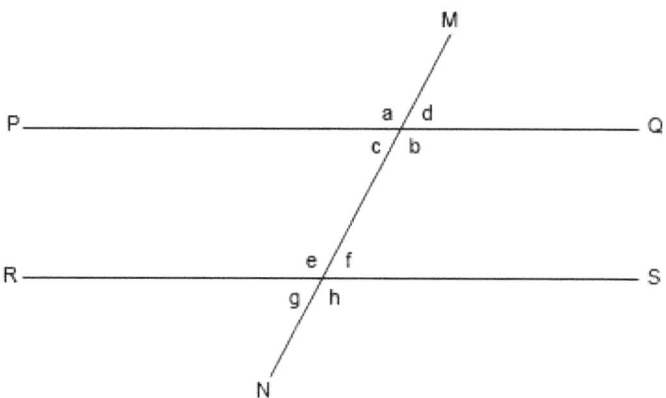

Fig. 2.4. Four pairs of vertically opposite angles.

A basic property of vertically opposite angles is that any angle in the first pair is *supplementary* to any angle in the second pair. *Supplementary* is the descriptive word that is used to refer to *a pair of angles if the sum of their measurements is equal to 180^0*. If the two supplementary angles are *adjacent* (that is, share a common vertex and a common side), then their non-shared sides form a straight line as shown in Fig. 2.5.

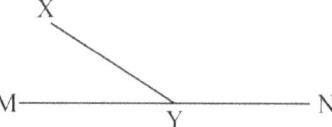

Fig. 2.5. $\angle$ MYX and $\angle$ NYX are Supplementary angles. $\angle$ MYX is adjacent to $\angle$ NYX.

The supplementary angles formed in Fig.2.5 are $\angle$ MYX and $\angle$ NYX (i.e. $\angle$ MYX $+ \angle$ NYX $= 180^0$). Their common vertex and common side are point Y and line XY, respectively.

Supplementary angles are usually contrasted with *complementary angles*. Two angles are considered *complementary* if the sum of their measurements is equal to 90^0. If the two *complementary angles* are adjacent, then their non-shared sides form a right angle, as in Fig. 2.6 (b). The two acute angles in a right triangle are complementary, since there are 180^0 in a triangle and 90 degrees are accounted for by the right angle. See Fig. 2.6 (a).

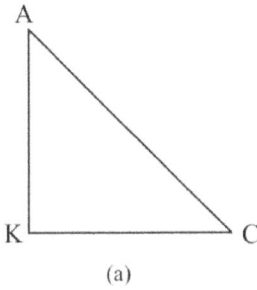 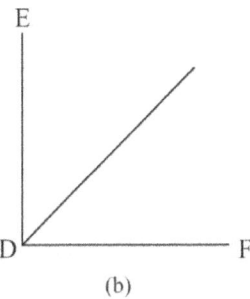

(a)   (b)

Fig. 2.6 (a) and (b). Complementary angles in a right triangle (a); adjacent complementary angles (b).

Rule 2: Lines are limitlessly extendable in both directions in a two-dimensional space such as the plane. All the lines we have described or drawn so far in this section invariably have a common characteristic which can be formulated as the following rule: *a pair of two different lines either are parallel to each other, or intersect each other at one and only one point. In other words, all lines in the same plane fall into two categories: those that are parallel to each other; and those that intersect each other at one point or another.*

In higher-dimensional planes, however, a pair of two lines may neither intersect each other, nor be parallel to each other. Two such lines are called *skew lines.*

Exercise 1: A drawing is shown in Fig. 2.7 in which line **ST** intersects line **QR** at point **T**. Line **YZ** also intersects line **QR** at point **Z**.

(a) What kinds of lines are ST, YZ, and QR?

(b) What is the relationship between ST and QR, between YZ and QR, between ST and YZ? Why do you think so?

(c) Name or describe at least two angles you can recognize in the drawing.

(d) What is the magnitude (degree measurement) of each such angle?

(e) Name or describe at least two angles in the drawing, each of which is equal to 180 degrees.

Solution to Exercise 1:

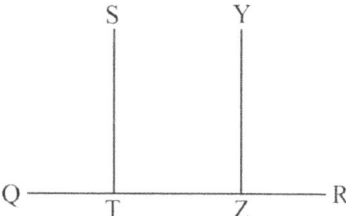

Fig. 2.7. Two vertical parallel lines ST and YZ are perpendicular to a horizontal line QR.

(a) Lines ST and YZ are both vertical lines, and line QR is a horizontal line.
(b) Both lines ST and YZ are perpendicular to line QR because each of them intersects QR at right angles (90^0). ST and YZ are parallel lines because they are separated by an equal distance at every point (i.e. they will never intersect or meet each other).
(c) Four recognizable angles in the drawing are: $\angle$QTS; $\angle$STZ; $\angle$YZT; and $\angle$YZR.
(d) The magnitude (or degree measurement) of each angle: $\angle$QTS = 90^0; $\angle$STZ = 90^0; $\angle$YZT = 90^0; and $\angle$YZR = 90^0.
(e) QT = ST = TZ = YZ = ZR = 180^0; QT, ST, TZ, YZ, and ZR are straight lines, also called open angles.

Rule 3: If lines **ST** and **YZ** are perpendicular to line **QR**, then lines **ST** and **YZ** are parallel to each other. Consequently, the following statement is true: *if two lines lying in the same plane are perpendicular to a third line, then the two lines are said to be parallel.*

Rule 4: In a rectangle, we can see that there are two pairs of parallel sides: one pair of horizontal parallel lines and another pair of vertical parallel lines. Therefore, the following statement is true: *if two lines lying in a plane have (or can have) a common point, then they are said to intersect each other; if they do not have (or cannot have) a common point, then they are said to be parallel to each other.*

However, there is yet another rare possibility to consider if two parallel lines have two or more common points. In such a case, we can say that the two parallel lines coincide.

Exercise 2: Let us refer back to Fig. 2.3 where line MN intersects line KL. Find the sizes of the other angles if $\angle MPK = 45^0$.

Solution to Exercise 2:

(a) $\angle MPK = 45^0$ – given;

(b) $\angle KPL = 180^0$ - a straight line or open angle;

(c) $\angle MPK$ is adjacent to $\angle MPL$, sharing a common side MP and a common vertex at point P. Accordingly, $\angle MPK + \angle MPL = \angle KPL = 180^0$.
(d) $\angle MPL = \angle KPL - \angle MPK = 180^0 - 45^0 = 135^0$;
(e) $\angle KPN = \angle MPL$ – vertically opposite angles are equal;
(f) $\angle NPL = \angle MPK = 45^0$ – vertically opposite angles are equal;

Exercise 3: A drawing is shown in Fig. 2.8. Lines **QR** and **ST** are perpendicular to each other. The point of intersection is **P**. Name and determine the sizes of the angles formed.

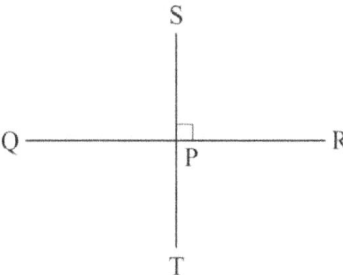

Fig. 2.8. Perpendicular lines QR and ST.

Solution to Exercise 3: Since lines **QR** and **ST** are perpendicular to each other, it means that they intersect each other at right angles; that is, all the four angles formed ($\angle SPR$, $\angle SPQ$, $\angle QPT$, and $\angle TPR$) are right angles: $\angle SPR = \angle SPQ = \angle QPT = \angle TPR = 90^0$. *It is easy to see that a right angle is equal to half of a straight angle (or open angle).*

Note: *Any straight line is a straight angle (or open angle). A straight angle or open angle is equal to 180^0. Two straight lines which intersect to form right angles are said to be perpendicular. In Fig.2.8, lines QR and ST are perpendicular. Usually, we write 'QR$\perp$ST' to indicate that 'line QR is perpendicular to line ST', or vice versa. Line segments lying on perpendicular lines are called perpendicular line segments.*

2. 3   Kinds of Quadrilaterals

Let us consider the following exercise.

Exercise 4:

    (a) **(i)** Draw an arbitrary twisted quadrilateral **EFHG** in which the opposite sides are parallel lines. Label it Fig.2.9.

       (ii) Name the two pairs of opposite sides of the quadrilateral **EFHG** in Fig. 2.9.

       (iii) Name or describe the angles formed in terms of their vertices.

       (iv) What is the name of such a quadrilateral, according to a definition   given from the beginning of this chapter?

    (b) **(i)** Draw an arbitrary quadrilateral **MNPO** of which two opposite sides are parallel and the other two opposite sides are non-parallel. Label it Fig.2.10.

       (ii) Name the two opposite sides that are parallel and the two opposite sides that are not parallel in Fig.2.10.

       (iii) What is the name of such a quadrilateral, according to a definition given from the beginning of this chapter?

Solution to Exercise 4:

    (a) **(i)** An arbitrary quadrilateral **EFHG** is represented in Fig.2.9.

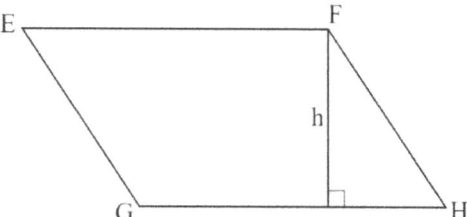

Fig. 2.9. Parallelogram EFHG.

       (ii) The two pairs of opposite sides of the quadrilateral **EFHG** in Fig. 2.9 are **EF** & **GH** and **EG** & **FH**. That is, **EF** $\parallel$ **GH** (**EF** is parallel to **GH**) and **EG** $\parallel$ **FH** (**EG** is parallel to **FH**).

       (iii) The angles formed are: $\angle$ **GEF**, or simply $\angle$ **E** (point **E** is the vertex of $\angle$ **GEF**); $\angle$ **EFH**, or simply $\angle$ **F** (point **F** is the vertex of $\angle$ **EFH**); $\angle$ **FHG**, or simply $\angle$ **H** (point **H** is the vertex of $\angle$ **FHG**); and $\angle$ **HGE**, or simply $\angle$ **G** (point **G** is the vertex of $\angle$**HGE**).

   (iv) The name of quadrilateral **EFHG** is *parallelogram.*
   A parallelogram is a quadrilateral having opposite parallel sides
   of equal length. The parallel sides of a parallelogram lie in
   parallel lines.

(b) **(i)** Another arbitrary quadrilateral **MNPO** is represented in Fig.2.10.

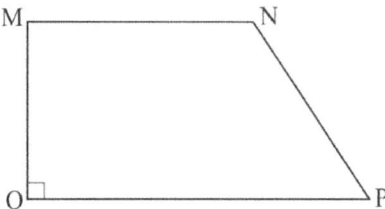

Fig. 2.10. Trapezoid MNPO.

   (ii) The two opposite parallel sides in Fig.2.10 are **MN** and **OP**;
   the two opposite non-parallel sides are **MO** and **NP**.

   (iii) The name of quadrilateral **MNPO** is *trapezoid.* A trapezoid is
   a quadrilateral having one or neither pair of sides parallel.
   A quadrilateral having two parallel sides of unequal length may
   also be referred to as a *trapezoid.*

Besides rectangles and squares, parallelograms and trapezoids are two common
types of quadrilaterals. The parallel sides of a trapezoid are called its ***bases***; and
the other two non-parallel sides are called its *lateral sides*.

We have said that a line joining any opposite vertices in a quadrilateral is called
a *diagonal*. There are only two diagonals that can be drawn in any given
quadrilateral. The diagonals that can be drawn in Fig. 2.9 are **EH** and **FG**, and
in Fig. 2.10 – **MP** and **NO**.

A line that is perpendicular to the opposite horizontal sides in a parallelogram is
called its *height*. The vertical line drawn from vertex **F** and perpendicular to the
base **GH** in Fig. 2.9 is the height of parallelogram **EFHG**. A line that is
perpendicular to the bases of a trapezoid is called the *height* of the trapezoid. In
Fig.2.10, **MO** represents the height of the trapezoid MNPO.

2.4　Areas of Triangle, Rectangle, Parallelogram, and Trapezoid

To find the area of a triangle or parallelogram means to determine how many square units are contained in the given figure. We can remember how we earlier in the 5th grade determined the area of a triangle by the formula $A = \frac{1}{2}lh$, where l is the length of one of the sides, and h is the height of the triangle, drawn from the vertex opposite l to side l. We also did determine the area of a rectangle using the formula $A = lw$, where l and w are the sides (*length* and *width*) of the rectangle.

There is a corresponding rule by which we can calculate the area of a parallelogram. In order to understand what this rule is about, let us consider the following exercises.

Exercise 5:

(a) (i) Draw a rectangle **OPQR** and label it Fig.2.11. Designate the middle between line **PQ** as point **S**, and the middle between **OR** as point **T**, such that line **ST** (draw perpendicular to **OR** and **PQ**) is the height of triangle **OSR** drawn within rectangle **OPQR**.

(ii) If **ST** = **PO** = **QR**, what is the area of triangle **OSR**?

(iii) If **PO** = **RQ** = 4 cm and **OR** = **PQ** = 5 cm, calculate the area of triangle **OSR**.

(b) Find the area of rectangle **OPQR**. What do you observe about the relationship between the areas of the rectangle and the triangle?

Solution to Exercise 5:

(a) (i)

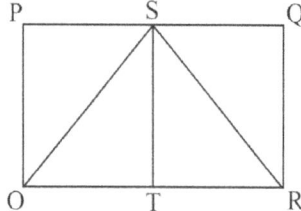

Fig. 2.11. Rectangle OPQR.

(ii) It is easy to be convinced that the area of the triangle **OSR** is equal to half of the product of its base **OR** and its height **ST**.

(iii) If **PO** = **RQ** = **ST** = 4 cm and **OR** = **PQ** = 5 cm, then the area of triangle **OSR** is:

$$A_{OSR} = \frac{1}{2}(OR \cdot ST) = \frac{1}{2}(5 \cdot 4) = 10 \text{ cm}^2 \text{ (square centimeters)}.$$

(b) The student is required to do it independently.

Exercise 6:

(a) Draw an isosceles triangle **KML** and label it Fig.2.12. The base of triangle **KML** is **KL** and the height is **MN**. **MN** is perpendicular to **KL**, dividing triangle **KML** into two equal right triangles **KMN** and **LMN**.

(b) (i) Transform triangle **KML** into a parallelogram, by drawing an adjacent triangle **MLO** which has or shares a common side **ML** with the original triangle **KML**. The base of triangle **MLO** is **MO**. Triangle **MLO** is inverted (upside down) but has the same dimensions as triangle **KML**. Label the transformation as Fig.2.13. It is not difficult to see that the resulting figure **KMOL** is a parallelogram.
(ii) Calculate the area of parallelogram **KMOL** which is represented in Fig. 2.13, if for example its height **MN** = **PL** = 6 cm, and its base **KL** = **MO** = 7 cm.

Solution to Exercise 6:

(a)

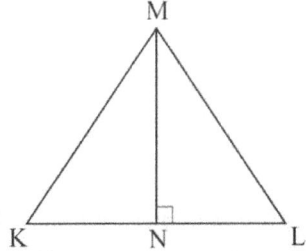

Fig.2.12. An isosceles triangle KML

(b) (i) Transformation of triangle **KML** into a parallelogram **KMOL** is shown below in Fig. 2.13.

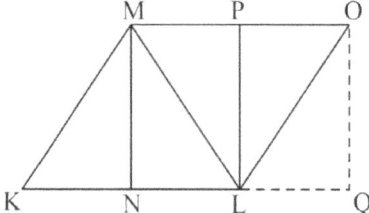

Fig.2.13. Comparison between a rectangle and parallelogram.

(ii) We can see that in the parallelogram **KMOL** (Fig. 2.13) the opposite sides are equal: **KL = MO**, and **MK = OL**. The height of triangle **KML** is equal to the height of triangle **MLO.** That is, **MN = PL**.

Moreover, From Fig.2.13, it can be seen how parallelogram **KMOL** can be transformed into a rectangle **MNOQ**. This means that the area of parallelogram **KMOL** is equal to the area of rectangle **MNOQ**. The base of parallelogram **KMOL** is **KL. KL = KN + NL = NL + LQ** (Fig.2.13). That is, the base of parallelogram **KMOL** is equal to the base of rectangle **MNOQ**. *Height* **MN** of rectangle **MNOQ** is also the *height* of parallelogram **KMOL**.

Accordingly, *the area of a parallelogram is equal to the product of its base and its height drawn to its base*: $A = bh$, Where b is the base, h is the height, and A is the area of the parallelogram.

Therefore, if $b = 7$ cm and $h = 6$ cm, the area A of parallelogram **KMOL** described in Fig. 2.13 is:

$$A_{kmol} = b \cdot h = 7 \cdot 6 = 42 \text{ cm}^2 \text{ (square centimetres),}$$

Where b is the base, h is the height, and A_{kmol} is the area of parallelogram **KMOL.**

Most of the geometric concepts handed down to us are attributed to Euclid. He is an outstanding Greek mathematician of 3rd century B.C. His famous work "Elements" set out the principles and system of geometry and remained a main text in geometry until the 19th century.

In order to set out the rule for finding the area of a trapezoid, it is expedient that we divide it into several parts such that from these parts it is possible to make up a figure (a triangle, rectangle, or parallelogram) area of which we can easily calculate, based on the preceding exercises. Let us consider a few more exercises to illustrate what we are talking about.

Exercise 7:

(a) (i) Draw a trapezoid **CDST** and label it Fig. 2.14. **DS** is the upper base and **CT** is the lower base of the trapezoid; **CT > DS**. It is necessary to divide the trapezoid into two parts so that from them it is possible to make up a triangle in each. For this purpose, draw a straight line (a diagonal) connecting vertices **D** and **T**. Now it can be seen that two triangles (△ **CDT** and △ **DST**) are formed (Fig.2.15). The symbol "△" is used to designate a "triangle".

(ii) Measure the base **CT** and height **DF** of △ **CDT** and calculate its area.

(b) If **CT** = 6 cm; **DF** = 3 cm; **DS** = 2 cm; **ST** = **CD** =3.5 cm; **DT** = 5 cm:

(i) Calculate the area of △ **CDT**.

(ii) Calculate approximately the area of trapezoid **CDST**

Solution to Exercise 7:

(a)   (i)

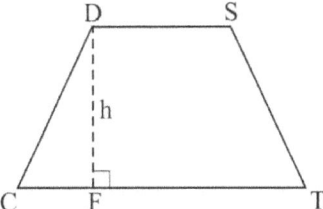

Fig.2.14. A trapezoid CDST.

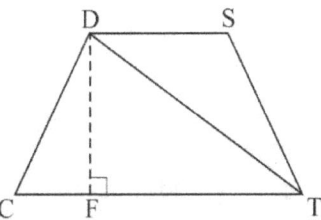

Fig.2.15. Division of trapezoid CDST into two parts forming two triangles
(△CDT and △DST).

(ii) The student is required to do it independently.

(b) (i) The base *b* of △ **CDT** is **CT** = 6 cm – given; and the height *h* of

$\triangle$ **CDT** is DF = 3 cm – given. The area $\mathbf{A_{cdt}}$ of $\triangle$ **CDT** is:

$$A_{cdt} = \frac{1}{2}(CT \cdot DF) = \frac{1}{2}(b \cdot h) = \frac{1}{2}(6 \cdot 3) =$$
$$= 9 \text{ cm}^2 \text{ (square centimetres)} - \text{the area of } \triangle \text{ CDT.}$$

(ii) *The area of trapezoid* ***CDST*** *is equal to the sum of the areas of* $\triangle$ ***CDT*** *and* $\triangle$ ***DST*** (Fig.2.15). In order to calculate the area of $\triangle$ **DST**, we ought to know its height; unfortunately, the height of $\triangle$ **DST** is not given. In order to determine its height, we should measure the length of a straight line drawn from vertex **S** to diagonal **DT**. It is important and required that the straight line must be perpendicular to **DT,** which is the base of $\triangle$ **DST**. If we assume that such a straight line is approximately equal to 1.2 cm, then the area $\mathbf{A_{dst}}$ of $\triangle$ **DST** is equal to the product of half of its base *b* and its height *h*:

$$A_{dst} = \frac{1}{2}(b \cdot h) = \frac{1}{2}(5 \cdot 1.2) = \frac{1}{2} \cdot 6 = 3 \text{ cm}^2 \text{ (square centimetres)} - \text{the area of}$$
$\triangle$ DST.

<u>PROOF</u>: As a means of proof that the above calculation is correct, we can otherwise determine the area of $\triangle$ **DST.** It is logical to suppose that the height of $\triangle$ **DST** is the same as that of $\triangle$ **CDT;** that is, the height *h* of $\triangle$ **DST** is **DF** = 3 cm. The base *b* of $\triangle$ **DST** (Fig. 2.15) is given above to be **DS** = 2 cm. Therefore, the area of $\triangle$ **DST** is:

$$A_{dst} = \frac{1}{2}(b \cdot h) = \frac{1}{2}(2 \cdot 3) = 3 \text{ cm}^2 \text{ (square centimetres)},$$
which is the same result we have obtained earlier.

Consequently, the area of trapezoid **CDST** may be determined as the sum of the areas of the two triangles ($\triangle$ **CDT** and $\triangle$ **DST**):

$$A_{cdst} = A_{cdt} + A_{dst} = 9 + 3 = 12 \text{ cm}^2 \text{ (square centimetres)},$$

where $\mathbf{A_{cdst}}$ is the area of trapezoid **CDST**, $\mathbf{A_{cdt}}$ is the area of triangle **CDT**, and $\mathbf{A_{dst}}$ is the area of triangle **DST**.

Calculation of the area of trapezoid **CDST** is possible to perform by means of other methods, one of which involves the following steps (See Fig. 2.16):

1. draw a perpendicular line from vertex **S** to the base **CT** so as to form a second right triangle;

2. designate the point under **S** as point **P,** and the second right triangle as △**SPT**;
3. determine the areas of right triangles **DFC** and **SPT**;
4. determine the area of the resulting rectangle **DSPF**; and
5. finally, add the areas of the two triangles and that of the rectangle to get the area of trapezoid **CDST**.

If the above steps are followed and performed correctly, the same result of 12 cm^2 (the area of trapezoid **CDST**) should be obtained.

Exercise 8:

(a) Calculate the area of trapezoid **CDST (Fig.2.16)** if: **CF** = **PT** = 2 cm; **DF** = **SP** = 3 cm; **DS** = **FP** = 2 cm.
(b) Compare the answers or results in Exercises 7 (b) and 8 (a).

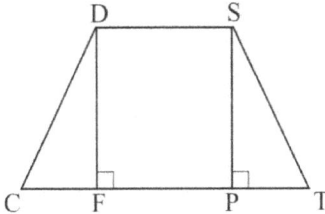

Fig. 2.16. Division of trapezoid CDST into two right triangles and a rectangle.

Solution to Exercise 8:

(a) The student is required to do it independently.
(b) The student is required to do it independently.

Exercise 9: Calculate the area of the trapezoid **DOPE** in Fig.2.17, if its bases DE = 9 cm, OP = 3 cm, and height h = OM = 4 cm.

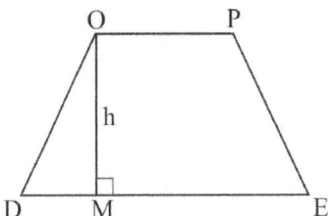

Fig.2.17. Trapezoid DOPE.

Solution to Exercise 9:

We divide trapezoid **DOPE** into a parallelogram (**NOPE**) and triangle (**DON**) as shown in Fig.2.18. The area of the trapezoid is equal to the sum of the areas of the parallelogram and triangle. NO = EP, NE = OP.

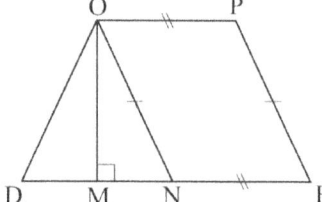

Fig. 2.18. Division of trapezoid DOPE into a parallelogram NOPE and a triangle DON.

We write a formula for calculating the area of the trapezoid **DOPE**:

$$A_{dope} = A_{nope} + A_{don} = (NE \cdot OM) + \frac{1}{2} (DN \cdot OM);$$

$$A_{dope} = 3 \cdot 4 + \frac{1}{2}(6 \cdot 4) = 12 + 12 = 24 \text{ cm}^2 - \text{the area of trapezoid } \textbf{DOPE},$$

where A_{nope} is the area of the parallelogram **NOPE**, A_{don} is the area of the triangle **DON**, **NE** is the base of the parallelogram **NOPE**, **OM** is the same height of both parallelogram **NOPE** and triangle **DON**, and **DN** is the base of the triangle **DON**.

Exercise 10: Find the area of a trapezoid **WXYZ** (Fig.2.19) having divided it by a diagonal into two triangles.

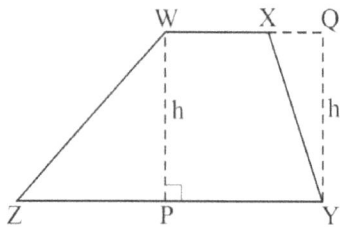

Fig.2.19. Trapezoid WXYZ.

Solution to Exercise 10:

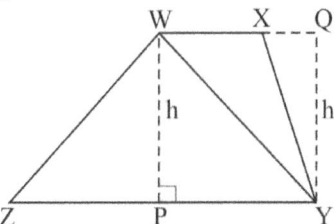

Fig.2.20. Division of trapezoid WXYZ by a diagonal into two triangles WYZ and WXY.

The area of the trapezoid **WXYZ** is equal to the sum of the areas of the two triangles (**WYZ** and **WXY**):

$$A_{wxyz} = A_{wyz} + A_{wxy}$$
$$= \frac{1}{2}(ZY \cdot WP) + \frac{1}{2}(WX \cdot QY).$$

The heights **WP** and **QY** of the two triangles are equal since each of them is equal to h, which is the height of the trapezoid **WXYZ**; that is **WP = QY = h**. **ZY** is the lower base, and **WX** is the upper base. Consequently,

$$\mathbf{A_{wxyz}} = \frac{1}{2}(ZY \cdot h) + \frac{1}{2}(WX \cdot h)$$
$$= \frac{1}{2}h \cdot (ZY + WX)$$
$$= \frac{1}{2}(b1 + b2) \cdot h,$$

Where A_{wxyz} is the area of the trapezoid **WXYZ**, A_{wyz} is the area of triangle WYZ, A_{wxy} is the area of triangle WXY, **b1** and **b2** are the bases of the two triangles (WYZ and WXY) making up the trapezoid **WXYZ**, and h is the height of the trapezoid **WXYZ**.

From here, we can conclude that *the area of a trapezoid is equal to half of the sum of its bases multiplied by its height:*

$$\mathbf{A_{trapezoid}} = \frac{1}{2}(\mathbf{b1} + \mathbf{b2}) \cdot \mathbf{h}$$

where $A_{\text{trapezoid}}$ is the area of the trapezoid, **b1** and **b2** are its bases, and **h** is its height.

2.5   Prism

If we consider the rectangular parallelepiped in Fig. 2.21, we will be reminded of its characteristic features. First of all, we will remember that every rectangular parallelepiped has six plane walls which are called its **borders**.

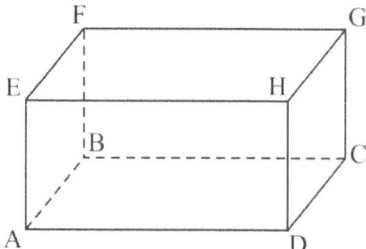

Fig.2.21. A rectangular parallelepiped, also regarded as a kind of prism.

Each border of a rectangular parallelepiped has the form of a rectangle. These borders intersect each other by straight lines which are called their *ribs* (or *edges*). All the edges of a rectangular parallelepiped intersect by three in one point which we already know as *vertices*. If we observe keenly the figure in Fig. 2.21, we can notice that the six borders of the rectangular parallelepiped represented are **ABCD**, **EFGH**, **AEHD**, **BFGC**, **AEFB**, and **DHGC**. It also has eight vertices which are **A, B, C, D, E, F, G**, and **H**; and twelve edges (**AB, BC, CD, AD, EF, FG, GH, EH, AE, BF, DH**, and **CG**). The height of the given rectangular parallelepiped is equal to **EA = FB = HD = GC**; the length is equal to **AD = EH = BC = FG**; and the width is equal to **EF = GH = AB = CD**.

Exercise 11:

(a) Draw an unfolded view of the rectangular parallelepiped in Fig. 2.21 and label it Fig. 2.22.

(b) Assuming that the length, width, and height of the rectangular parallelepiped in Fig. 2.21 are 4 cm, 1.5 cm, and 2 cm, respectively, calculate: **(i)** the area of its lateral surface
    (ii) the area of its total surface, and
    (iii) its volume.

Solution to Exercise 11:

(a) An unfolded view of the rectangular parallelepiped in Fig. 2.21 is shown in Fig. 2.22 as required.

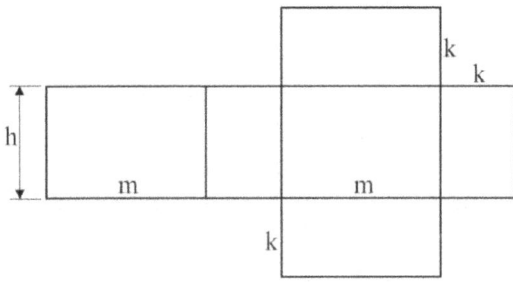

Fig.2.22. An unfolded view of the rectangular parallelepiped in Fig. 2.21.

(b)
(i) Let's analyze the configuration in Fig. 2.21 in order to be able to derive the formula by which we can calculate both areas of the lateral and total surfaces of a rectangular parallelepiped as required in the problem.

As stated above, every rectangular parallelepiped has six plane borders, which are in essence its surfaces. *The area of the lateral surface of the rectangular parallelepiped in Fig. 2.21 is the area of the front (AEHD), rear (BFGC), left and right (AEFB + DHGC) borders.* Since each border of a rectangular parallelepiped has a rectangular form, it is obvious that the area of AEHD is also equal to that of BFGC, and that the area of AEFB is also equal to that of DHGC. *Accordingly, the dimensions (length l, width w, height h) of the front border (AEHD) is equal to those of the rear border (BFGC), and that the dimensions of the left border (AEFB) is also equal to those of the right border (DHGC).*

Consequently, based on the above analysis, we can calculate the area of the lateral surface of a rectangular parallelepiped:

$$A_{ls} = A_{aehd} + A_{bfgc} + A_{aefb} + A_{dhgc};$$
$$A_{aehd} = A_{bfgc} = lh;$$
$$A_{aefb} = A_{dhgc} = hw;$$

$$A_{ls} = lh + lh + hw + hw$$
$$= 2(lh) + 2(hw)$$
$$= 2h \cdot (l + w)$$
$$A_{ls} = 2 (l + w) \cdot h$$
$$A_{ls} = p \cdot h;$$

$$\boxed{A_{ls} = p \cdot h}$$

where A_{ls} is the area of the lateral surface of a given parallelepiped whose base is a rectangle; in such a case, $p = 2 \cdot (l + w)$ is the perimeter of its rectangular base; h is its height; l is its length; and w is its width.

NOTE: The formula $(A_{ls} = p \cdot h)$ indicates that the area of the lateral surface of a rectangular parallelepiped is equal to the product of the perimeter of its base and its height. *If the base of the given parallelepiped is a square, then $p = 4a$ (where a is the length of the side of the square).*

(ii) The area of the total surface of a rectangular parallelepiped is equal to the sum of the area of its lateral surface and the area of its top *(EFGH)* and bottom *(ABCD)* borders. It should also be recognized that *the dimensions of the top border (EFGH) is equal to those of the bottom border (ABCD); it follows from here, therefore, that the area of EFGH is likewise equal to that of ABCD.*

NOTE: The area of the *top* and *bottom* borders of a rectangular parallelepiped is usually referred to as "the area of its bases", and may be designated as:

$$A_b + A_b = 2A_b = lw + lw = 2(lw).$$

Strictly speaking, the area of a base of a rectangular parallelepiped designated by A_b is equal to $l \cdot w = lw$ (i.e. the product of the *length* and the *width*). Since there are two bases (the *top* and *bottom* borders), they are added up to be $2A_b$, as reflected in the formula below.

Hence the area of the total surface of a rectangular parallelepiped may be calculated by the formula:

$$A_{ts} = A_{ls} + 2A_b;$$
$$A_b = A_{abcd} + A_{efgh};$$
$$\text{Since } A_{abcd} = A_{efgh} = lw,$$
$$\text{then } 2A_b = lw + lw = 2(lw);$$
$$\therefore A_{ts} = A_{ls} + 2A_b$$
$$= A_{ls} + 2(lw);$$

$$A_{ts} = A_{ls} + 2A_b = (p \cdot h) + 2(lw)$$

where A_{ts} is the area of the total surface of a rectangular parallelepiped, A_{ls} = $(p \cdot h)$ is the area of the lateral surface of the rectangular parallelepiped, and $2A_b = 2(lw)$ is the area of its top and bottom bases.

(iii) *The volume of a rectangular parallelepiped is equal to the product of its three dimensions (length l, width w, and height h):*

$$V_{rp} = l \cdot w \cdot h.$$

Exercise 12:

(a) Draw a cube and label it Fig. 2.23.
(b) **(i)** Draw an unfolded view of the cube in Fig. 2.23 and label it Fig. 2.24;
   (ii) Find the area of the total surface of the cube in Fig. 2.23, if the length of its edges is 5 cm.
   (iii) Find the volume of the cube in Fig. 2.23.

Solution to Exercise 12:

(a) In Fig.2.23, a cube is drawn. It should be noted that a cube is a solid figure having six plane square faces or surfaces in which the angle between two adjacent surfaces is a right angle.

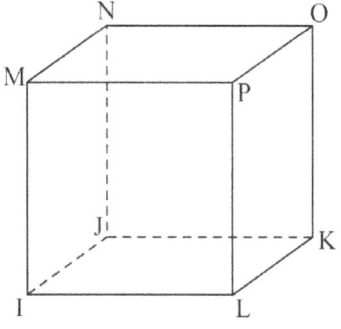

Fig.2.23. A cube.

(b) **(i)** An unfolded view of the cube in Fig. 2.23 is shown in Fig. 2.24 as required.

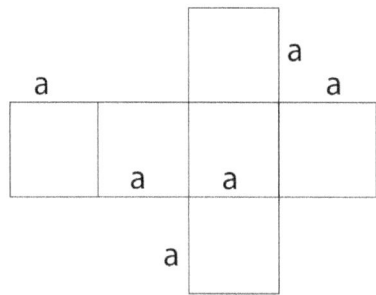

Fig.2.24. An unfolded view of a cube.

(ii) The area of the total surface of a cube is equal to the sum of the areas of its six plane surfaces or borders. The dimensions (length l, width w, and height h) of a cube as in Fig.2.23 are equal, i.e. $l = w = h = a = 5$ cm; hence the area of its total surface may be calculated by the expression:

$$\begin{aligned}
A_{ts\text{-cube}} &= 2(l \cdot h) + 2(h \cdot w) + 2(l \cdot w); \\
&= 2(a \cdot a) + 2(a \cdot a) + 2(a \cdot a) \\
&= 2a^2 + 2a^2 + 2a^2 \\
&= 6a^2
\end{aligned}$$

$$A_{\text{ts-cube}} = 6a^2$$

where $A_{\text{ts-cube}}$ is the area of the total surface of a cube. In the given problem, where $a = 5$ cm,

$$A_{\text{ts-cube}} = 6a^2 = 6 \cdot 5^2 = 6 \cdot 25 = 150 \text{ cm}^2.$$

(iii) The volume of a cube is the product of its three dimensions (length l, width w, and height h). The volume of the cube in Fig.2.23 is given by the expression:

$$V_{cube} = lwh;$$

Since $l = w = h = a$, the above expression becomes:

$$V_{cube} = a \cdot a \cdot a = a^3$$

In the given problem, where $a = 5$ cm,

$$V_{cube} = a^3 = 5^3 = 125 \text{ cm}^3.$$

With the help of the preceding exercises, we have set forth the methods and formulae by which it is possible to calculate the surface areas of rectangular parallelepipeds and cubes, as well as their respective volumes.

In mathematics, a *rectangular* parallelepiped is also referred to as a *prism*. Rectangular parallelepipeds and cubes are types of *rectangular prism* (also may be called *right prism*). If the base of a prism is made up of a triangle, then such a prism is said to be a *triangular prism*. A triangular prism is shown in Fig. 2.25. All the three lateral plane surfaces or borders of a triangular prism are rectangles. The other two borders at the top and bottom are equal triangles. It is customary to refer to these triangles as the *bases* of the prism, and the other rectangular surfaces as *lateral borders*. In other words, the lateral borders of a right prism, in essence, are rectangles. The base may be made up of polygonal (multi-angular) figure with different number of sides.

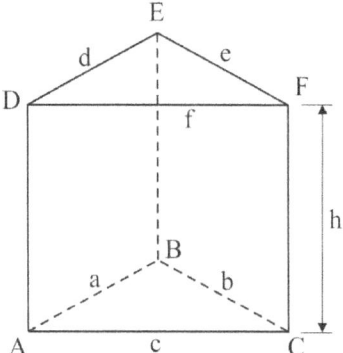

Fig.2.25 A triangular prism.

All prisms possess the following characteristics: every prism has two equal borders which do not intersect each other (these are the top and bottom surfaces, or bases); all other borders or plane surfaces are rectangles (these are the lateral borders). See Fig.2.26. Accordingly, the names of prisms are determined by the number of angles of the polygon which constitutes or makes up the base. For example, the base may be a triangle, quadrilateral, etc. It is easy to recognize that a *triangular prism* has three lateral borders; a *quadrilateral prism* has four lateral borders; a *pentagonal prism* has five lateral borders; a *hexagonal prism* has six lateral borders; and so on. Fig. 2.26 below is an example of a hexagonal prism.

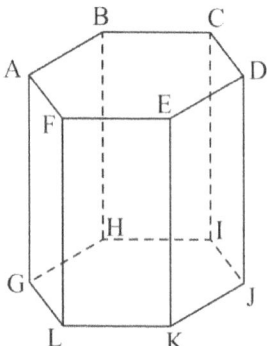

Fig.2.26. A hexagonal prism.

The area of the lateral surface of a right triangular prism (see Fig. 2.25) is equal to the perimeter of its base multiplied by its height (that is, the product of *the perimeter of its base* and *its height*):

$$A_{tp} = (a + b + c) \cdot h$$
$$= (d + e + f) \cdot h$$
$$= p \cdot h,$$

$$\boxed{A_{tp} = p \cdot h}$$

where A_{tp} is the area of the lateral surface of the right triangular prism; $p = (a + b + c) = (d + e + f)$ is the *perimeter* of its base; a, b, and c, as well as d, e, and f are the respective sides of the triangles at the bottom and top bases of the prism; and h is the *height* of the prism.

NOTE: *To find the area of the total surface of a right triangular prism, the areas of the triangles at the top and bottom surfaces (i.e. the area of its bases) should be added to the area of its lateral surface as determined above. That is, the area of the total surface of a right triangular prism is the sum of the areas of its bases (its top and bottom surfaces) and its lateral surface.*

The Volume of a Right Prism

To calculate the volume of a right prism, the formula for finding the volume of a rectangular parallelepiped may be written in the form of:

$$V_{rp} = Ah,$$

$$\boxed{V_{rp} = Ah}$$

where V_{rp} is the volume of the right prism; $A = lw$ is the area of the base, and h is the height of the prism.

2.6   Pyramids

Let us look at the figure represented in Fig. 2.27 below. Such figures are referred to as *pyramids*. Take note of the number of borders, edges, and vertices of which a pyramid is made of. As we can see in Fig. 2.27, the lateral borders in all pyramids are triangles. The bases of pyramids may differ by form. The base of a pyramid may be a triangle, a quadrilateral, a hexagon, etc. The base of the pyramid in Fig. 2.27 (a) is a triangle. A pyramid with a triangular base is called a *triangular pyramid*; one with a quadrilateral base is called a *quadrilateral pyramid*; one with a hexagonal base is called a *hexagonal pyramid*, and so on. Example of a triangular pyramid is shown in Fig. 2.27 (a). An unfolded view of the triangular pyramid in Fig. 2.27 (a) is represented in Fig. 2.27 (b). Its base is made up of an equilateral triangle.

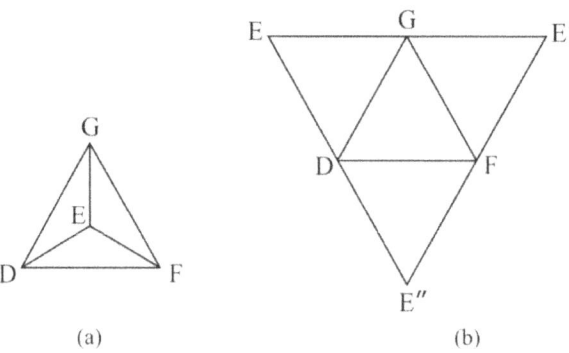

(a)   (b)

Fig. 2.27. (a) A triangular pyramid and (b) its unfolded view.

If we compare a pyramid with a prism, we find that the latter has two bases, and that the former has only one base. The lateral borders of a prism are rectangles, while the lateral borders of a pyramid are triangles. The lateral edges (or ribs) of a prism do not intersect each other; in a pyramid all lateral edges intersect or converge at a point at the top, which is called the vertex of the pyramid.

The surface of every pyramid consists of triangles (making up the lateral borders) and a polygon (forming the base of the pyramid). The area of the lateral surface of a pyramid may be calculated with the help of the formula:

$$A_{ls\text{-}pyramid} = A_1 + A_2 + ...+ A_n$$

where $A_{ls\text{-}pyramid}$ is the area of the lateral surface of the pyramid; A_1, A_2, and A_n represent respective border areas of the lateral surfaces; and n indicates the number of lateral borders.

The area of the total surface of a pyramid may be determined by the formula:

$$A_{\text{ts-pyramid}} = A_{\text{ls-pyramid}} + A_b$$

where $A_{ts\text{-}pyramid}$ is the area of the total surface of the pyramid; $A_{ls\text{-}pyramid}$ is the area of its lateral surface, and A_b is the area of its base.

The volume of any pyramid is equal to one-third of the product of the area of its base and its height:

$$V_{\text{pyramid}} = \frac{1}{3}(A_b \cdot h)$$

where $V_{pyramid}$ is the *volume of the given pyramid,* A_b is the area of its base, *and* **h** *is its height.*

The latest formula shows that *the volume of a pyramid is three times less than the volume of a prism with correspondingly equal base and height.* This conclusion was made by the 5th century B.C. ancient Greek philosopher Democritus. He discovered that *the volume of a pyramid is equal to a third of the volume of a prism with the same base and height.* The proof of this conclusion was also given by another ancient Greek mathematician in the 6th century.

Chapter Two: Questions to Test Your Understanding

1. Define the following terms:

(a) Perpendicular lines
(b) Parallel lines
(c) Rectangle
(d) Trapezoid
(e) Parallelogram
(f) Polyhedron
(g) Point

(h) Line
(i) Angle

2. Draw the following:

(a) Perpendicular lines
(b) Parallel lines
(c) Parallelogram
(d) Trapezoid

3. Name few objects which have the form of a rectangle, parallelogram, and trapezoid.

4. How the area of a parallelogram is calculated?

5. How the area of a trapezoid is calculated?

6. Draw a model of a right prism. Measure its dimensions and calculate the area of its lateral surface and its volume.

7. Draw a model of a right triangular prism: (a) Measure its dimensions: (b) Calculate the area of its lateral surface; (c) Calculate its volume.

8. Draw a model of a triangular pyramid: (a) Measure its dimensions: (b) Calculate the area of its lateral surface; (c) Calculate its volume.

9. Draw a model of a quadrilateral pyramid: (a) Measure its dimensions: (b) Calculate the areas of its lateral and full surfaces; (c) Calculate its volume.

10. Write the formula by which the area of each of the following figures may be calculated: (a)   Rectangle (b) Triangle (c) Parallelogram (d) Trapezoid

11. Write the formula by which the area of each of the following figures may be calculated:
    (a) Prism (b) Pyramid.

Chapter Two: Problems and Exercises

1. With the help of a drawing triangle or protractor, draw a straight line MN through point F (Fig. 2.28) such that MN is perpendicular to the straight line ST.

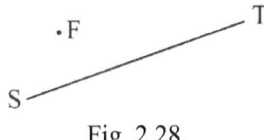

Fig. 2.28.

2. In Fig. 2.29, MN $\perp$ OP, $\angle$ RCT = 120^0, and $\angle$ PCT = 30^0. Calculate $\angle$ RCO and $\angle$ TCO.

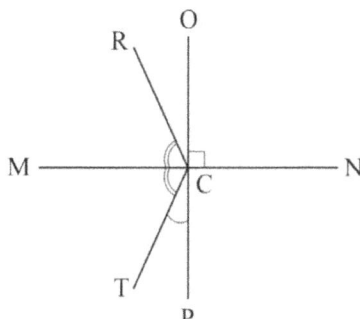

Fig. 2.29. To calculate $\angle$ RCO and $\angle$ TCO.

3. Two taxi cabs traveled by straight-line routes from points C and D, respectively, to a train station designated by point P (Fig. 2.30). The distance CP which the first taxi traveled is 5 miles, and the distance DP which the second taxi traveled is 8 miles. What is the approximate distance between points C and D?

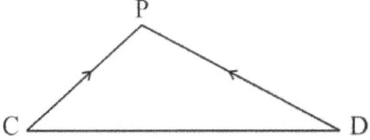

Fig. 2.30. To determine the approximate distance between points C and D.

4. Explain which of the following statements are true and which are not.

(a) If DE$\perp$GH and GH $\perp$ XY, then DE $\perp$XY.
(b) If straight line AB$\perp$straight line CD, then CD$\perp$AB.
(c) Two perpendicular straight lines on a plane always intersect each other.
(d) Two straight lines which intersect each other are always perpendicular
(e) Each of the angles formed by the intersection of two perpendicular lines is always equal to 90^0.

(f) A straight line regardless of its length is equal to 180^0.

5. In Fig. 2.31 (in textbook), $XY \perp WZ$, $\angle RTX + \angle STZ = 55^0$, and $\angle YTS = 70^0$. Calculate $\angle RTX$, $\angle RTZ$, and $\angle XTS$.

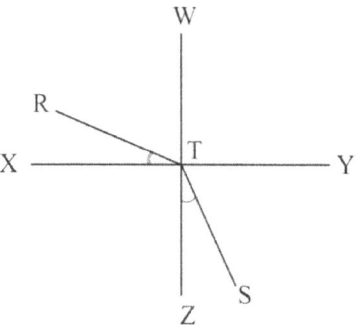

Fig. 2.31. To calculate $\angle RTX$, $\angle RTZ$, and $\angle XTS$.

6. A parallelogram MNOP is shown Fig.2.32. Compare the diagonals in the figure. Which diagonal of the parallelogram is greater? Does a parallelogram exist in which the diagonals are equal? If yes, then how is it called? Answer the questions, using direct measurements of the dimensions.

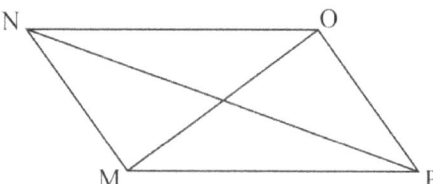

Fig. 2.32. To compare the diagonals in the figure.

7. In parallelogram EFGH (Fig. 2.33), EH = 7 cm, GH = 5cm. What are the measurements of EF and FG? Calculate the parallelogram's perimeter and area, if its height h is equal to 3 cm.

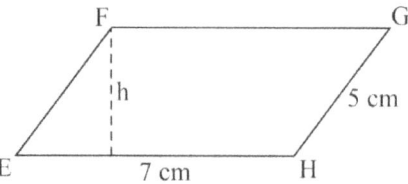

Fig. 2.33. To calculate the parallelogram's perimeter and area.

8. Two sides of a given parallelogram RSTU are in the ratio of 6:5. Its perimeter is equal to 66 cm. Find the dimensions of its sides.

9. The perimeter of a parallelogram BCDE (Fig. 2.34) is equal to 20 cm. Find the length of diagonal CE, if the perimeter of $\triangle$BCE is equal to 16 cm.

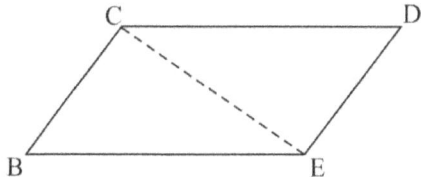

Fig. 2.34. To find the length of diagonal CE.

10. Draw a general model of a parallelogram, and tell whether or not its area changes depending on the angle between its sides. Explain why.

11. Draw two parallelograms with the same base 12 cm which have the same area 48 cm². What is the height of these parallelograms?

12. Take the necessary measurements of the figure, plan of which is represented in Fig. 2.35, and calculate its actual area. The scale used in the plan of the figure is 1:1,000.

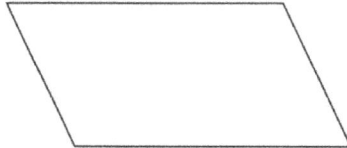

Fig. 2.35. To calculate the actual area of the figure.

13. A rectangle with sides of 4 cm and 3 cm; a parallelogram with sides of 4 cm and 3 cm; and a square with side of 3.5 cm. Which of these polygons has the greatest area? Compare their perimeters.

14. A field has the form of a rectangle, the base b and height h of which are 700 m and 150 m, respectively. A five-lane coal tar highway (15 m in width) passes through the field under a right angle (that is, perpendicular) to its base.
   (a) Draw a sketch of the field.
   (b) Find the total area of the field.
   (c) What is the area of the portion of the highway covered by the field?

(d) What is the remaining area of the field (apart from the area of its portion through which the highway passes)?

15. How does the area of a square change, if its side increases by four times?

16. By how many time it is necessary to decrease the side of a square so that its area is decreased by 9 times?

17. Determine the sides of a rectangle, if they are in the ratio of 4:5, and its perimeter is equal to 270 cm. Find the area of the rectangle.

18. A rectangle and a parallelogram have identical bases and areas. How do their perimeters compare? Draw the figure to support or justify your answer.

19. Calculate the area of a trapezoid if its bases are 9 cm and 14 cm, and its height is 7 cm.

20. Find the total area of the figure having a form as shown in Fig. 2.36 and having the following dimensions: AB = 2.5 cm; BC = 2 cm; DE = 2 cm; height h = 3.5 cm; AF = 7 cm.

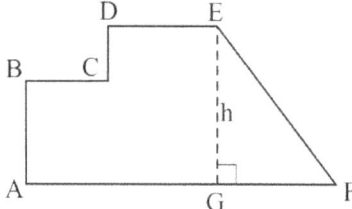

Fig. 2.36. To find the total area of the figure.

21. Calculate the area of a trapezoid if one of its bases is equal to 27 cm, and the other base is 1.5 times greater than the first. The height of the trapezoid is 25 % of the larger base.

22. How many kilograms of seeds are required in order to sow a farm land having the form of a trapezoid? The larger base of the trapezoid is 2.5 km; its smaller base is three-fifths of the larger, and its height is equal to three-fourths of the smaller base. *The quantity of seeds required to sow one hectare of the farm land is 150 kg.*

23. In Fig. 2.37, the plan for a plot of land is shown. The scale used in the sketch is 1:500; and the dimensions are indicated in millimeters. Calculate the area of the given plot, if ST = 400 mm.

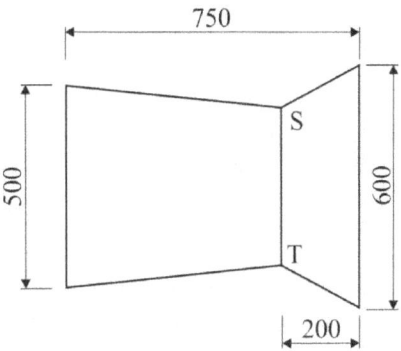

Fig. 2.37. To calculate the area of the plot.

24. Draw a rectangular parallelepiped. Designate the number of *vertices*, *borders*, and *edges*, using capital letters. Determine the area of each border and the volume of the figure you have drawn.

25. The dimensions of a rectangular parallelepiped are 7 cm, 11 cm, and 9 cm. Draw this figure and calculate the following:
   (a) the area of its lateral surface;
   (b) the area of its total surface;
   (c) its volume.

26. Cut from a paper an unfolded model of a right prism, having a triangular base. Two of the sides of the triangular base are each 2.2 cm, and the third side is 3 cm. The height of the prism is 3.5 cm. Do the following exercises:
   (a) Glue together a prism from this unfolded model.
   (b) Measure the height of the triangular base of the prism.
   (c) Calculate the area of the triangular base of the prism.
   (d) Calculate the area of the total surface of this prism.

27. The base of a triangular prism is a right triangle with cathetuses (the base and height) of 3 cm and 4 cm. The height of the prism is 7 cm. Calculate the area of the surface and the volume of this prism. *A cathetus in a right triangle is either of the sides other than the hypotenuse.*

28. The base of a quadrilateral prism is a parallelogram with sides 9 cm and 10 cm. The height of the parallelogram is equal to 5 cm, and the height of the prism is 12 cm. Calculate the area of the total surface and the volume of the prism.

29. Perform the necessary measurements of a given triangular pyramid and calculate the area of the total surface and the volume.

30. Draw an unfolded model of a quadrilateral pyramid, the base of which is a rectangle having the sides 3 cm and 4 cm. All lateral edges of the pyramid have the same length. Glue together a model of the pyramid from the unfolded form and calculate its total surface. The height of each triangle making up the lateral surface of the pyramid is 6 cm.

31. Calculate the volume of a quadrilateral pyramid, the base of which is a square whose side is 5 cm. The height of the pyramid is 9 cm.

Chapter Two: Exercises to Rate Your Ability

1. Draw a prism and perform the necessary measurements of its dimensions. Calculate: (a) the area of its lateral surface; (b) the area of its total surface; and (c) its volume.

2. A parallelogram has sides of 4 cm and 3 cm. Find the perimeter and area of this parallelogram, if its height is 2 cm.

3. Draw three parallelograms with base 5 cm and height 3.5 cm. Which of these parallelograms has the greatest perimeter?

4. Draw a triangular prism and carry out the necessary measurements of its dimensions. Calculate: (a) the area of its lateral surface; (b) the area of its total surface; and (c) its volume.

5. The bases of a trapezoid are 7 cm and 6 cm. Its height is 3 cm. Find the area of this trapezoid. Measure its lateral sides and calculate its perimeter

6. Calculate the volume of a quadrilateral pyramid having a rectangular base, the sides of which are 13 cm and 15 cm. The height of the pyramid is 10 cm.

7. Draw and measure the dimensions of an unfolded form of a triangular pyramid whose base is an equilateral triangle. All lateral edges of the pyramid have identical length. The height of each triangle making up the lateral surface of the pyramid measures 4 cm. Calculate: (a) the area of the lateral surface; and (b) the area of the total surface of the pyramid.

8. The dimensions of a rectangular parallelepiped are 2 cm, 3 cm and 4 cm. Calculate: (a) the area of its lateral surface; (b) the area of its base; (c) the area of its total surface; and (d) its volume.

9. In Fig. 2.38, MN⊥OP, ∠CBO + ∠OBN = 140^0, and ∠ABN = 65^0. Calculate ∠MBC and ∠PBA.

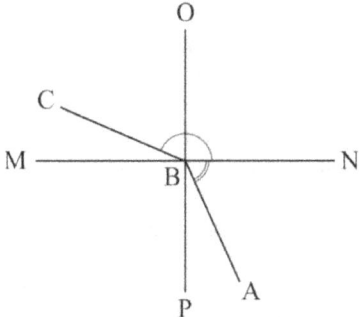

Fig. 2.38. To calculate ∠MBC and ∠PBA.

Chapter Three: Circular and Spherical Bodies

This chapter is dedicated to discussions about circular bodies, including the length of circumference, surface areas of circles, cylinders, cones, as well as spherical solids. We have already discussed concepts about circumference and the area of a circle in previous sections. However, they are further treated here, since they form part and parcel of concepts concerning circular bodies.

3.1 The Length of a Circumference and the Area of a Circle

Circumference is a geometric category or figure consisting of all points on a plane which are equidistant from a given point considered to be the center of the circumference. A circumference strictly describes a circle. A **circle** is the part of a plane entirely enclosed by a circumference. In other words, a *circumference* is the measurement of the length around a circle, that is, the perimeter of the circle. Accordingly, the terms "circumference" and "circle" may be used interchangeably to suggest the same meaning or concept. A circle, as a circumference, has a *center*, *radius*, *diameter*, and *chords*. A circle is represented in Fig. 3.1 in which point **C** is its center; **AC** = **BC** = **DC** is its radius; **BD** is its diameter; **EF** is a chord; **AD** and **AB** are *arcs*, respectively. Besides that, separate parts of a circle have special names. For example, the part of a circle enclosed by two radii (the plural form of the word *radius*) and an arc is called a *sector*. **ABC** in Fig. 3.1 is a sector of the circle shown. The part of a circle enclosed by a chord and an arc is called a *segment*.

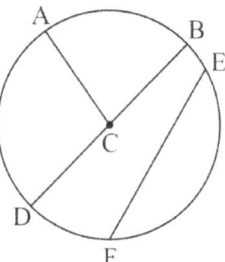

Fig. 3.1. A circle.

Exercise 1:

(a) Measure the length of the circumference *c* of any circular object (a coin, the bottom or brim of a drinking cup, for example) and its diameter *d* with the help of either a ribbon or a thread. Use a measuring tape, if it is available. Repeat the measurement about three times. Record the results of the first, second, and third measurements in a table.

(b) Determine the ratio of the length of the circumference c of the circular object to its diameter d, and record it in the table after each measurement. (The table shall have four columns with headings **measurement, c, d,** and **c : d.** Under the heading **measurement** shall be **first, second,** and **third.**)

If the exercise is performed correctly, you will discover that *the length of the circumference c of the circular object is approximately 3.14 times greater than the length of its diameter d.* In other words:

$$c = 3.14d.$$

Since we know that $d = 2r$, we can rewrite the expression of the circumference as

$$c = 3.14 \ (2r) = 6.28r.$$

From here, it can be seen that the length of the circumference c of a circle is 6.28 times greater than the length of its radius r.

It has been proven that the ratio of the length of a circumference to its diameter is expressed by an infinite decimal fraction:

$$c \div d = 3.142856...$$

This ratio of the length of a circumference of a circle to its diameter is designated by a Greek letter π (called *pi*). Therefore, the formula of the length of a circumference of a circle to its diameter may be written as:

$$c = \pi d.$$

For practical calculations, it is accepted to take the number π as **3.14**, or the fraction **22 ÷ 7**. The ancient Greek mathematician *Archimedes* (287 – 212 B.C) was the first to calculate the value of the number π. He gave the value of the number π to be between $3\frac{10}{71}$ and $3\frac{1}{7}$, that is, $3\frac{10}{71} < \pi < 3\frac{1}{7}$.

Either in the 5th grade or somewhere in a previous section of this book, we talked about circle and circumference. There we gave the formula for calculating the area of a circle πr^2 without any discussion as to how the formula was derived. Here we will look very briefly at some details concerning the rule which governs this formula.

In order to establish the rule for calculating the area of a circle, it is expedient and important first of all to think about the parameters of a circle that factor into the calculation of its area: its radius r, its diameter d, and its perimeter or circumference c. Interestingly, it is well known that the calculations of the areas of most polygons (such as triangle, rectangle, parallelogram, and trapezoid) involve the multiplication of the *base* (in some cases, the *length*) and the *height*. This fact gives us a clue to think of the *radius* of a circle as its *height*, since it is the *length from the center of the circle to its surface*. Consequently, *to determine the base of a circle*, we can divide it into several sectors. In our particular case here, we have divided a circle into eight sectors, which are made up of *curvilinear* (curved) figure as shown in Fig. 3.2 (b).

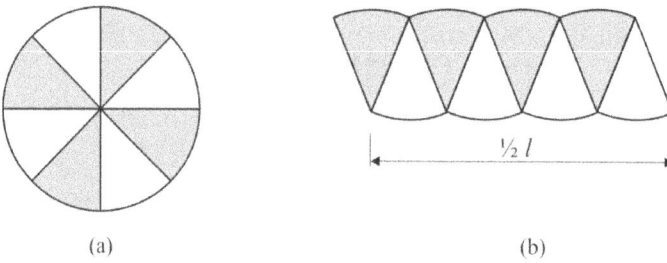

(a)   (b)

Fig. 3.2 (a) and (b). Calculation of the area of a circle

The drawing in Fig. 3.2 (b) by form resembles a parallelogram. The difference between them is that in the figure represented in Fig. 3.2 (b) the two opposite sides are not *rectilinear* or straightforward; they consist of curvilinear links. It is not difficult to observe that the greater the number of sectors into which the circle is divided, the closer the represented figure approaches or matches the form of a parallelogram. The *base* of such a parallelogram *is equal to half of the length of the circumference* of the circle. In this way, knowing the *base* (which is half the circumference) and the *height* (which is the radius), we can establish the rule for calculating the area of the circle. Consequently, according to the foregoing analysis, *the area of a circle is equal to the product of half the length of its circumference and its radius*:

$$
\begin{aligned}
A_{circle} &= \tfrac{1}{2}\,c \cdot r \\
&= \tfrac{1}{2}\,(\pi d)r \\
&= \tfrac{1}{2} \cdot (2\pi r) \cdot r \\
&= \pi r^2;
\end{aligned}
$$

$$A_{circle} = \pi r^2$$

where A_{circle} is the area of the circle; c is its circumference; d is its diameter; and r is its radius. It should be noted that in Fig. 3.2 (b) above, l is equal to the circumference c (the measure of the length of the perimeter of the circle). That is, $\frac{1}{2}l = \frac{1}{2}c$, as applied in the above formula to determine the area of the circle.

3.2   Cylinders

A rounded metallic fish or vegetable can, a pipe or tube, and many other such similar objects do have the form of a *cylinder*. A cylinder is actually a solid figure consisting of two identical circles interconnected at every point by a set of parallel lines which are usually perpendicular to the circular planes. The two circles form the upper and lower bases of the cylinder, and the parallel lines make up its lateral surface. A typical cylindrical form is represented in Fig. 3.3 (a). What other objects do you know, having the form of a cylinder?

Fig. 3.3 (b) depicts a cylinder, which is formed when the rectangle QRST rotates around one of its sides. Side ST in Fig. 3.3 (b) is called the *generatrix*. It describes the lateral surface of the cylinder, since during rotation of rectangle QRST around QR, it forms a cylinder. Side QT in the lower base of the cylinder describes a circle with radius QT = RS (in the upper base of the cylinder). Side QR (= ST) is called the *height* of the cylinder.

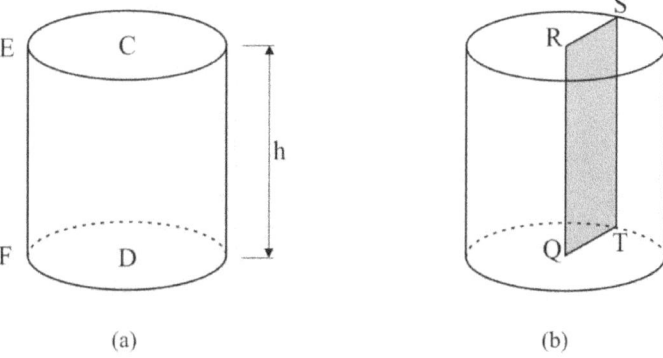

(a)   (b)

Fig. 3.3. (a) A cylinder; (b) Formation of a cylinder.

In Fig. 3.3 (a), **C** and **D** represent the two circles which are said to be the bases of the cylinder. **EF** represents the height h of its lateral surface; that is, EF = h.

Exercise 2:

(a) Carefully wind or wrap a paper around an object in the form of a cylinder, observing that the paper exactly fits the form and size of the lateral surface of the object.

(b) Unwrap or unfold the paper around the object in the cylindrical form, and carefully observe the form and size of the paper.

(c) What do you notice about the form and size of the paper?

(d) Note and be convinced that the circumference c of the circular base of the cylinder is equal to QR = TS = $2\pi r$, and also that the radius r of its circular base is equal to QT = RS (see Fig. 3.3 (b)).

The form and size of the paper tell us that if we unfold the lateral surface of a cylinder, it perfectly iterates or exemplifies the form of a rectangle. The length of the base of the rectangle is equal to the length (or perimeter) of the circumference c of each base of the cylinder (c = QR = ST = $2\pi r$). Thus, we can calculate the area of the lateral surface of the cylinder with the help of the size of the paper which is in a rectangular form. This may be done by multiplying the length of the *base* of the rectangle by its *height*. The height of the rectangle is equal to the height of the cylinder QT = CD = RS = h. We can likewise calculate the areas of the two circles forming the upper and lower bases of the cylinder, since we know their circumference. Consequently, we can finally determine the area of the total surface of the cylinder, which is the sum of the areas of its lateral surface and its circular bases. Fig. 3.4 represents an unfolded view of a cylinder, which consists of a rectangle and two circles.

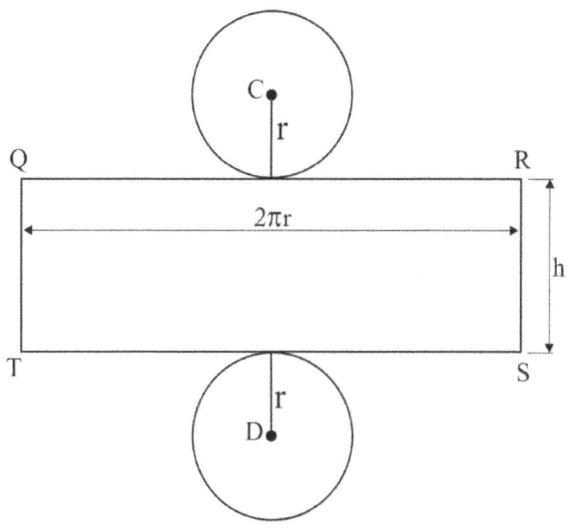

Fig. 3.4. An unfolded view of the cylinder in Fig. 3.3 (a).

Based on the above analysis, the area of the lateral surface of a cylinder is equal to the area of the rectangle, the base and height of which are equal to $2\pi r$ and h, respectively. In other words, *the area of the lateral surface of a cylinder is equal to the product of its base (which is the circumference of a circle) and its height:*

$$A_{\text{ls-cylinder}} = c \cdot h$$
$$= 2\pi r \cdot h$$
$$= 2\pi rh,$$

$$\boxed{A_{\text{ls-cylinder}} = 2\pi rh}$$

where $A_{\text{ls-cylinder}}$ **is** the area of the lateral surface of a cylinder; $c = 2\pi r$ is the base of the cylinder (which is the circumference of a circle); and h is the height of the cylinder.

The area of the total surface of a cylinder is the sum of the areas of its lateral surface and its two circular bases:

$$A_{\text{ts-cylinder}} = A_{\text{ls-cylinder}} + A_{\text{b-cylinder}}$$
$$= 2\pi rh + (\pi r^2 + \pi r^2),$$

87

$$A_{ts\text{-}cylinder} = 2\pi rh + 2\pi r^2$$

where $A_{ts\text{-}cylinder}$ is area of the total surface of a cylinder.

The calculation of the volume of a cylinder is similar to the calculation of the volume of a prism by the formula:

$$V_{cylinder} = Ah$$
$$= \pi r^2 h,$$

$$V_{cylinder} = \pi r^2 h$$

Where $V_{cylinder}$ is the volume of the cylinder; $A = \pi r^2$ is the area of its circular base; and h is its height.

3.3   Cones

A *cone* is a solid figure consisting of a plane base bounded usually by a circle or ellipse, every point of which is joined to a fixed point. The vertex of a cone lies outside the plane of the base. A cone may be formed by rotating a right triangle **FPE** around one of its cathetuses as in Fig.3.5. Cathetus **EP**, in this case, is the *height h* of the cone. The second Cathetus **FP** describes a circle which is said to be the *base* of the cone. Line segment **EF** is referred to as the *generatrix* of the cone. A *generatrix* is a point, line, or plane that is moved in a specific way to produce a geometric figure.

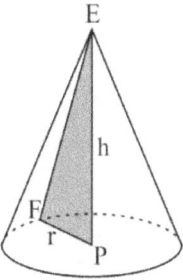

Fig. 3.5. A cone.

Exercise 3:

    (a) Carefully wrap a paper around a cone or an object in the form of a cone so as to capture or reproduce the form of a come. Observe that the paper exactly fit the form and size of the cone.

    (b) Think over the definition of the *lateral and total surface* of a cone.

    (c) Unfold the paper around the cone or the cone-like object, and carefully observe the form and size of the paper.

    (d) What do you notice about the form and size of the paper? *Carefully compare the length of the circumference of the base of the cone with the length of the unfolded arc of the lateral surface of the cone, after which you should be convinced that they are equal.*

If we unfold the paper we have wrapped around the cone or cone-like object, then its unfolded form would appear as in Fig. 3.6. With the help of the unfolded form of the cone, we are able to determine its *lateral* and *total surface*. This is possible because *the length of the circumference* **c** *of the circular base of the cone is equal to the length of the arc of the unfolded form of the lateral surface of the cone.* In other words, the length of the circumference c of the cone's circular base is equal to the length of the arc of the unfolded paper representing the cone's lateral surface. As we can see in Fig. 3.6, the length of the arc of the unfolded form of the cone's lateral surface is, in essence, the circle's circumference c; and it is equal to $c = 2\pi r$.

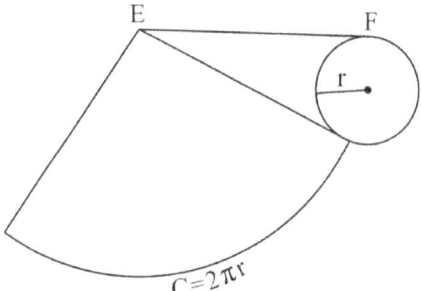

Fig. 3.6. An unfolded view of a cone.

Let us take a look at Fig. 3. 7 below. We can see that the unfolded form of the cone consists of a *sector* (divided into smaller unit sectors), a *radius* EF_0, and a *circle*, lying in the base of the cone. It can be seen in the figure that the radius of the sector is $\mathbf{EF_0} = l$ and that the circle makes up the base of the cone. (Actually, the *radius* of the circular base f of the cone is proportional to the *length* of the triangular form). The area of the lateral surface of the cone consists of smaller unit sectors $\mathbf{F_0EF_1}$, $\mathbf{F_1EF_2}$, $\mathbf{F_2EF_3}$, ..., $\mathbf{F_{n-1}EF_n}$. Each smaller unit sector has the form of a triangle. $\mathbf{A_1}$, $\mathbf{A_2}$, and so on to $\mathbf{A_n}$ are the areas of the respective smaller unit sectors which make up the larger triangle-shaped sector. Therefore, the area of the lateral surface of the cone is equal to the sum of the areas of these smaller unit sectors:

$\mathbf{A_{ls\text{-}cone}} = \mathbf{A_1} + \mathbf{A_2} + ... + \mathbf{A_n}$ - the area of the lateral surface of the cone.

$c = 2\pi r = f_1 + f_2 + ... + f_n$ - the length of the circumference of the circular base of the cone.

Since each unit sector has the form of a triangle, we can rewrite the expression for $\mathbf{A_{ls\text{-}cone}}$ as the following:

$$A_{ls\text{-}cone} = \tfrac{1}{2} \cdot l \cdot f_1 + \tfrac{1}{2} \cdot l \cdot f_2 + ... + \tfrac{1}{2} \cdot l \cdot f_n$$
$$= \tfrac{1}{2} \cdot l \cdot (f_1 + f_2 + ... + f_n)$$
$$= \tfrac{1}{2} \cdot l \cdot c$$
$$= \tfrac{1}{2} \cdot l \cdot (2\pi r)$$
$$= \tfrac{1}{2} \cdot 2 \cdot l \cdot (\pi r)$$
$$= \pi r l,$$

$$A_{ls\text{-}cone} = \pi r l$$

In the above formula, $A_{ls\text{-}cone}$ is the area of the lateral surface of the cone; $c = 2\pi r$ is the length of the circumference of the cone's base; r is the radius of the cone's base; f_1, f_2, f_n are the sectorial units into which the length of the circumference has been divided; and l is the length of the cone's lateral surface.

The area of the total surface of a cone is equal to the sum of the area of its lateral surface plus the area of its circular base:

$$A_{ts\text{-}cone} = A_{ls\text{-}cone} + A_{base\text{-}cone}$$
$$= \pi r l + \pi r^2$$

$$A_{ts\text{-}cone} = \pi r l + \pi r^2$$

In the above formula, $A_{ts\text{-}cone}$ is the area of the total surface of the cone; and $A_{base\text{-}cone}$ $(=\pi r^2)$ is the area of its circular base.

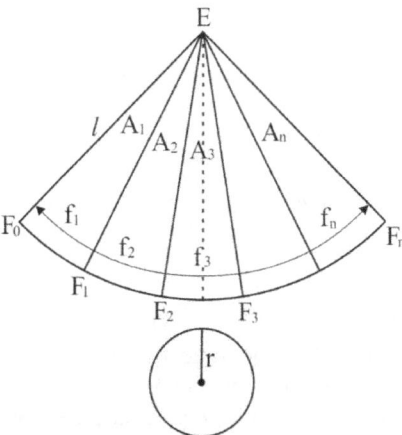

Fig. 3.7. Component parts of an unfolded view of a cone. $A_{ls} = \pi r l$.

If you compare the form of a cone with that of a pyramid, you will notice some similarities and some distinctions. Explain some similarities and some distinctions between the two figures.

The calculation of the volume of a cone is similar to the calculation of the volume of a pyramid by the formula:

$$V_{cone} = \frac{1}{3} A_b h$$

$$= \frac{1}{3} \pi r^2 h;$$

$$V_{cone} = \frac{1}{3} \pi r^2 h = \frac{1}{3} A_b h$$

A_b is the area of the circular base of the cone; h is the height of the cone; and r is the radius of the cone's circular base.

The formula for the calculation of a cone's volume, as given above, means that *the volume of a cone is equal to one-third of the volume of a cylinder with the same area of the circular base and the same height.* It is possible to be convinced in the correctness of the above formula, pouring sand from a hollow cone into a cylinder having the same base and height.

3.4   Spherical Solids

Objects such as a globe, tennis ball, basketball, and ball bearings relate to what we refer to as *spherical solids*. Each one of us has probably seen one kind or another of these objects and knows of their spherical form. Some fruits such as water melon, apples, cherries, and peas have form which resembles spherical solids. Strictly speaking, all of these possess the form of a *sphere*.

The moon, Earth, and other planets and celestial bodies possess form which resembles that of a sphere. It is possible to get a clear imagination about a sphere, rotating a semi-circle around its diameter. Let us take a look at Fig. 3.8. The line segment connecting two points on the surface of the sphere and passing through its center is called the *diameter* of the sphere. It is important to note that a circle is formed by any plane in any section of a sphere.

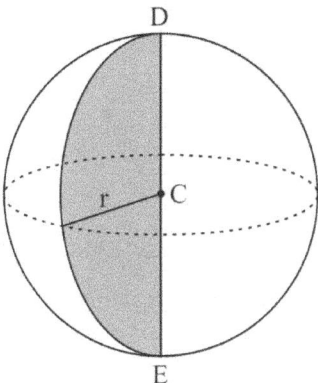

Fig. 3.8. A sphere.

If a secant plane passes through the center of a sphere, then a circle is formed in that section. The radius of such a circle is equal to the radius of the sphere. Such a circle is called a *great circle*. The circumferences of great circles on the globe, for example, are the equator and the meridians.

The lateral surfaces of a cone and a cylinder are curved surfaces; but it is possible to transform them into plane surfaces by way of *'straightening or unbending'* them; that is, it is possible to place them on a plane surface like a table to unfold their forms so as to be able to calculate the areas of their surfaces and their volumes as in the previous exercises (see Figs. 3.4, 3.6, and 3.7).

However, the surface of a sphere does not lend itself to *'straightening or unbending'*. In other words, the surface of a sphere does not possess the right characteristics or qualities for straightening it out into a plane surface. Consequently, it is not possible at this level to directly derive the formula for calculating the area of the surface of a sphere, using the previous *method of unfolding*. Thus, the following formulae for calculating the area of the surface of a sphere and its volume, until in upper mathematics classes, cannot be derived or proven at this stage.

The area of the surface of a sphere is equal to four times the area of a great circle:

$$A_{sphere} = 4A_{circle}$$
$$= 4\pi r^2,$$

$$A_{sphere} = 4\pi r^2$$

where A_{sphere} is the *area of the surface of a sphere, and* **r** *is the radius of the sphere.*

The volume of a sphere may be calculated by the formula:

$$V_{sphere} = \frac{4}{3}(\pi r^2)$$

Chapter Three: Questions to Test Your Understanding

1. Define the following terms:
 (a) Diameter
 (b) Radius
 (c) Chord of a circle
 (d) Segment of a circle

2. What is the distinction between a circumference and a circle?

3. Write the formula by which the following may be calculated:
 (a) The length or perimeter of a circumference
 (b) The area of a circle
 (c) The area of the total surface of a cylinder
 (d) The volume of a cylinder
 (e) The volume of a cone
 (f) The surface area of a cone
 (g) The surface area of a sphere

4. What is the value of the number π?

5. By how many times the length of a given circumference is more than its diameter?

6. Name a few objects which have the form of a: (a) cylinder (b) cone (c) sphere.

7. By which figure it is possible to exchange with the lateral surface of a cylinder?

8. Draw a cylinder and its unfolded form.

9. Compare and contrast the form of a cone and that of a pyramid.

10. Compare and contrast the form of a cone and that of a cylinder.

Chapter Three: Problems and Exercises

1. Find the circumference of a circle if its radius is 5 cm. Draw the circle and measure its diameter.

2. Find the diameter of a circle if its radius is 3.5 cm.

3. Draw a circle with diameter of 4 cm. Designate the center of the circumference by a capital letter. Name the following: all points belonging to the circumference and circle; all points belonging to the circle; all points not belonging to the circumference; all points not belonging to the circumference and to the circle.

4. Draw a circumference radius of which is equal to 30 cm. Draw two diameters that are perpendicular, and determine the length of the greatest chord.

5. The diameter of the circumference of a circle is designated MN. The greatest distance from a given point P to the circumference is equal to 25 mm; and the smallest distance from point P to the circumference is 15 mm. What is the length of the radius of this circumference? Examine possible cases of P (within and outside the circle).

6. Assuming that the radius of the Earth's equator is about 6,378 km. What is the perimeter (or length) of the Earth's equator, if π = 3.14. Round off your answer to the nearest unit.

7. Calculate the diameter of Mars, if the circumference of the planet is 21,308.64 km. Round off your answer to the nearest hundredth.

8. A person standing on a plane (flat) surface can see around himself at a distance of 1.5 km. What is the length of the circumference of the horizon which the person sees?

9. What distance can the end of the minute hand of a wall clock go for 12 hours, if the length of the minute hand is 6.5 cm?

10. A load is lifted or raised with the help of a pulley. At what distance is the load lifted for 10 rotations (or turns) of the pulley, diameter of which is 25 cm? *A pulley is a lever or wheel with a grooved rim in which a rope, chain, or belt can run in order to change the direction or point of application of a force applied to the rope, etc..*

11. The diameter of the shaft of a well is 15 cm, and the depth of the well is 12 m. How many times is it necessary to turn or rotate the handle of the shaft in order to draw out a bucket of water from the well? Round off your answer to the nearest unit.

12. The length of a circumference is 205 cm. By how much in length the radius of the circumference would increase, if the length of the circumference was increased by 1.5 m?

13. The diameter of a circumference was decreased by 2.5 cm. By how much the length of the circumference would decrease?

14. Moving with a speed of 350 mph, calculate the time for which it is possible to fly in an airplane around the Earth along the equator, if it is assumed that the radius of the equator is approximately 3,729 miles.

15. Calculate the area of a circle having the radius of 30 cm.

16. The external diameter of a pipe is 7 cm. The thickness of the walls of the pipe is 3.5 mm. Calculate the area of the ring of a section of this pipe (the cross-sectional area, in other words). Round off your answer to the nearest hundredth.

17. Carry out the necessary measurements and calculate the area of the shaded figure in Fig. 3.9.

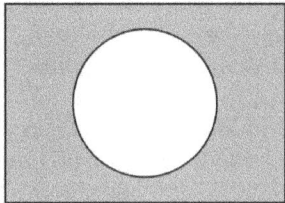

Fig. 3.9. To find the area of the shaded part in a figure.

18. The circumference of the complex of a football stadium is 35 miles. What is the area of this complex?

19. (a) Draw two circumferences such that the radius of one is three times greater than the radius of the other.
(b) Calculate how many times the length of one of the circumferences is greater than the length of the other.
(c) Calculate how many times the area of one circle is greater than the area of the other.

20. What is the area of a circle (rounded off to the nearest thousandth), if the length of its circumference is equal to 18.75 cm?

21. Calculate the area of the gasket shown in Fig. 3.10. The measurements are given in inches. Round off your answer to the nearest tenth. *A gasket is a compressible packing piece of paper, rubber, asbestos, etc., sandwiched between the faces of a metal joint to provide a seal.*

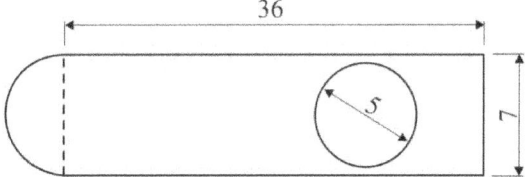

Fig. 3.10. To find the area of the gasket.

22. The length and width of a piece of paper in the form of a rectangle are 90 cm and 50 cm, respectively. It is necessary to cut from this piece of paper a circle of the largest possible size. What is the area of such a circle? Calculate what percentage of the rectangular piece of paper is wasted as a result of cutting the circle from it?

23. It is necessary to replace two water pipes (each with a diameter of 32 cm) with one new water pipe of the same water-carrying capacity. What would be the diameter of the new water pipe?

24. An aluminum wire with diameter 9 mm was stretched such that the new cross-sectional area became two times less than the original cross-sectional area. Determine the new dimensions of this aluminum wire, i. e. the new diameter and the new cross-sectional area of the stretched aluminum wire.

25. Draw an unfolded view of a cylinder, diameter and height of which are 18 cm and 6 cm, respectively. Find: (a) the area of its lateral surface; (b) the area of its base; (c) the area of its total surface; and (d) its volume.

26. Calculate the volume and the area of the total surface of a cone if the generatrix of the cone is 35 cm, the radius of the cone's base is 15 cm, and the cone's height is 25 cm.

27. Calculate the volume and the area of a sphere having a radius of 27 cm.

28. A lathe operator (metal turner) turned two ball-bearings from a metallic material. The diameters of the two ball-bearings are 2.5 in and 1.5 in, respectively. About how many times the mass of the larger ball-bearing is greater than that of the smaller?

29. Two pineapples having a spherical form are being sold by a market woman. One of the pineapples is 1.8 times larger than the other and two times more expensive. Which of the pineapples is cheaper and more advantageous to buy?

30. Calculate the volume of a sphere, if the length of its great circle is 92.5 in?

31. Assuming that the Earth has the form of a sphere with equatorial diameter of 12,756 km, calculate the area of the surface of the Earth and its equatorial circumference.

Chapter Three: Exercises to Rate Your Ability

1. Calculate the area of a circle having a radius of 15 cm.

2. Calculate the area of the total surface and the volume of a cylinder, radius of the base of which is 11 cm. The cylinder's height is 5.5 cm.

3. Calculate the area of the surface and the volume of a sphere, diameter of which is 18.5 in.

4. Carry out the necessary measurements and calculate the following:

   (a) The volume and the area of the surface of an orange, a tennis ball, or any other available circular object.
   (b) The area and the volume of a given cylinder.
   (c) The circumference and the cross-sectional area of an orange, a tennis ball, or any other available circular object.

5. Calculate the area of the total surface and the volume of a cone, diameter of the base of which is equal to 9.5 dm. The cone's height and generatrix are 2.5 dm and 4.5 dm, respectively.

6. Calculate the area of the shaded part of the circle shown in Fig. 3.11, if the diameter of the larger circle is 38 mm and that of the smaller circle is 17 mm.

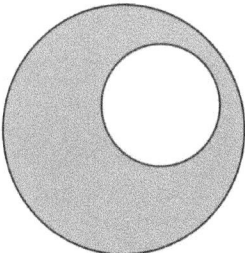

Fig. 3.11. To calculate the area of the shaded part of the circle.

Chapter Four

Aptitude Tests: Problems and Exercises for the Entire Course of Principles of Modern Elementary Mathematics

<u>Aptitude Test One</u>

Perform the indicated operations:

1. $(-2\frac{2}{5}) + (-3)^2 \cdot 2;$

2. $(-3 + 1\frac{1}{5})^2 (-5 + 4\frac{1}{3});$

3. $(-18.25) \div (-3.4 + 7.5 - 2.6)^2;$

4. $(-2 + \frac{1}{5})^2 (3 - \frac{3}{4});$

5. $(-\frac{3}{7} - \frac{4}{7})^2 + (-\frac{2}{7}) - 2(-\frac{5}{7})^2;$

6. $5(-0.03) \div (-2) - 7.5(-3);$

7. $3 \div (\frac{0.5 - 2}{6})^2;$

8. $\frac{0.5}{0.2 + 0.6} + (-\frac{1}{8});$

9. $-9 - (-12) \div 4 - 7.2;$

10. $(-5)^2 \cdot 15 - (-3)^2 \cdot 6^2;$

11. $(-2\frac{3}{8} + 5)(-1\frac{1}{4})^2;$

12. $(-4) - 24 \div 0.8 - 16.5;$

13. $-15 \div (-1.2 + 3\frac{1}{2} - 0.8);$

14. $(-2 + \frac{3}{5})^2 \div (2 - \frac{1}{5})^2;$

15. $(3 - \frac{4}{9})(4 - 2\frac{3}{4});$

16. $-3\frac{3}{4} + 6 \div (-1.2) + 9 - 5.2 \cdot 3^2$;

17. $6.45 - (-8.10) \div 9 + (-4.50 + 3.40) \div (-2.75)$;

18. $(3\frac{3}{5} + (-36.18 \div 9) - (-2.36) - 2.72) \div (-2\frac{1}{4})^2$;

19. $7 \div (- (-\frac{1}{8}) \div (- 0.75) + \frac{2}{9}) + 5\frac{1}{6}$;

20. $10 - (-5) \div (0.5) + 0.8 - 3.7 \div 4$;

21. $(-\frac{1}{5} \cdot 0.25) \div (-5) + \frac{2}{5} - (\frac{-1}{9}) \div (-2)^2$;

Open the brackets and simplify the following expressions:

22. $(3\frac{5}{6}x - 4\frac{2}{3}) - \frac{1}{9}$;

23. $-7x\,(-5 + 6x)$;

24. $(-3\frac{1}{8}a + \frac{3}{8}) - a(\frac{3}{4} - \frac{7}{8})$;

25. $-2a + 3b \cdot (4a - 7) - 12a$;

26. $9x\,(-3x + 4)$;

27. $\frac{3}{5} \cdot (\frac{5}{6}a - 4) - 7 \cdot (3\frac{3}{4}a)$;

Find the value of the given expression:

28. $y - 5 \cdot (2y - 3)$, if $y = 2^2$;

29. $5ab - 7.3ab$, if $a = 5$ and $b = 2\frac{1}{4}$;

30. $5\frac{3}{4}x - 3y$, if $x = - 0.4$ and $y = 1\frac{1}{5}$;

Aptitude Test Two

Find the value of the given expression:

1. $|-5.1| + |-4.7| - |8.3|$;

2. $\frac{4}{5} - |-3.5| + 5.6$;

3. $(\frac{3}{5})^2 - (\frac{4}{5})^2 - 8$;

4. $(-5)^2 + 7^2 - 9^2 - 32$;

5. $|-6.3| + |4.2| \cdot (-3)^2$;

Open the brackets and simplify the given expressions:

6. $(\frac{3}{13} - \frac{2}{9} + \frac{4}{5}) - (\frac{5}{26} - \frac{7}{18} - \frac{3}{25})$;

7. $(7a - 4) - (5a - 6)$;

8. $(-11x - 9y + 5) - (-8x + 12y - 9)$;

9. $(5.5a - 3.3b + 4.4) - (3.3a + 4.4b - 5.5)$;

10. $(-\frac{5}{7}x - \frac{4}{9}y + \frac{1}{3}) - (\frac{3}{7}x - \frac{1}{9}y + 2)$;

Find the value of the given expression:

11. $-6\frac{2}{3}x \cdot (-0.35y) \cdot \frac{1}{4}$, if $x = 3^2$ and $y = 4^2$;

12. $-3.5(-\frac{1}{6}x) \cdot (-3\frac{1}{3}y)$, if $x = (\frac{1}{4})^2$ and $y = (-2)^3$;

Solve the given equations:

13. $4\frac{3}{7} - \frac{5}{6} = x$;

14. $0.65 - (8 + a) = 15$;

15. $7 - (x + 3) = 18$;

16. $-9x + 3 = 5x - 12$;

102

17. $\dfrac{y-5}{3} = 1.8$;

18. $5.3 = \dfrac{20}{a-5}$;

19. $\dfrac{b}{2} = \dfrac{3+b}{12}$;

20. $2.5 \div 3 = x \div 5$;

21. $26 - 5 \cdot (6.5 - 4x) = 7$;

22. $\dfrac{-4}{|y|} = -0.03$;

23. $5\,|x| = 0.75$;

24. $|6x| = \dfrac{7}{36}$;

25. $-\dfrac{4}{5}\,|a| + 4.2 = -19$;

26. $-4 \cdot (2a + 8) + 5 = 7 - 9 \cdot (a + 3)$;

27. $6x - 3 \cdot (x + 10) = 4x - 2 \cdot (x + 6)$;

28. $0.8(2y - 5) - 3 = 0.12(y - 4) - 0.2$;

29. $12.4 - 15a + 5a = 0.6a - 20$;

30. $-\dfrac{4}{11} \div y = -3\dfrac{1}{22}$;

Aptitude Test Three

Solve the following equations:

1. $\dfrac{k}{-0.5} = 9$;

2. $\dfrac{1.5}{2n} = -3.6$;

Calculate the value of the given expression:

3. $(2x - 3) \div (4 - 5x)$, if $x = (-\dfrac{2}{3})^2$;

4. $|x \div y|$, if $x = -3$ and $y = 2$;

5. $|x| \div |y|$, if $x = (-2)^3$ and $y = (-5)^2$;

Which of the two expressions is greater, and by how much greater?

6. $(-\frac{4}{5})^3$ or $(-\frac{5}{6})^2$;

7. $(-3.2)^3$ or $(-0.32)^2$;

Which of the two expressions is greater, and by how many times greater? (*Note:* It is required to compare only the modules of the two numbers).

8. $(-\frac{3}{4})^2$ Or $(-\frac{2}{5})^3$;

9. $(-\frac{1}{2})^2$ or $(-\frac{2}{3})^3$;

10. Subtract the product of the numbers $-5\frac{3}{5}$ and $-6\frac{1}{4}$ from the sum of the numbers $\frac{5}{6}$ and $\frac{3}{5}$.

11. Subtract the sum of the numbers -4.9 and 7.5 from the product of the numbers $3\frac{1}{2}$ and $2\frac{3}{5}$.

12. Subtract the difference of 5.9 and -2.3 from the sum of 10.11 and -15.2.

13. Add 6.5 to the quotient gained from dividing -8.1 by 3.

14. By how much the difference of $2\frac{3}{4}$ and $\frac{7}{8}$ is less or more than the quotient gained from dividing $-15\frac{1}{2}$ by $3\frac{3}{4}$?

15. Add the product of -7 and $-\frac{4}{9}$ to the difference of 6.2 and $3\frac{2}{5}$.

16. Add the product of 11 and -7 to the quotient of $(-3)^2$ and $-\frac{5}{6}$.

17. 12 percent of $150,000 is equal to what?

18. What is 8.5% of 75 miles?

19. 22% of a number is equal to 88. Find the number.

20. Find a number if 0.6% of it is equal to 7.2.

21. 6 minutes is what percent of one hour?

22. 24 m is what percent of 600 m?

23. 0.6 hour is what percent of 4.8 hours?

24. Perform the indicated operations: $(5 - 2\frac{2}{3} \cdot 2.4 \cdot (-\frac{5}{4})^2) \div 1\frac{1}{2}$.

Solve the equation:

25. $\frac{2}{3} + (0.5 - x) = \frac{1}{2}$.

26. $0.4 \div y = 2.3 \div 5$.

27. Find the product of the roots of the equation $|x| = 9$.

28. Find the value of the expression: $-1.2 + 3 \cdot (-2\frac{1}{3}x^2 + 5)$, if $x = -2$.

Solve the equations:

29. $-2.2 \cdot (1 + \frac{10}{11}x) = 0.6x + 0.9$.

30. $4a \div \frac{1}{2} = 2\frac{1}{3} \div 5.2$.

Aptitude Test Four

Solve the equations:

1. $2\frac{3}{4}d = \frac{3}{8}$.

2. $3d \div 2\frac{1}{2} = \frac{7}{8}$.

3. What was the speed of a bus if it traveled 156 miles for $2\frac{1}{6}$ hours?

4. The mass of one cubic decimeter (i.e. 1 dm^3) of an iron is equal to $8\frac{2}{5}$ kg. What is the mass of $\frac{3}{8}$ dm^3 of this iron?

5. Calculate the mass of a layer of rock having length of 95 m, width of 35 m and thickness of $2\frac{1}{2}$ m, if $5\frac{3}{4}$ m^3 of the rock has a mass of 750 kg.

6. For $\frac{3}{5}$ of an hour a taxi cab traveled 36 km. What distance would this taxi cab travel for $2\frac{2}{3}$ hours under the same rate of motion?

7. Mr. Sumo walked 7.4 miles with a speed of 3 mph from his farm to a near-by village. Calculate the time he spent on his journey between his farm and the village.

8. The base of a reservoir completely filled with water is a rectangle with sides $9\frac{3}{5}$ meters and $15\frac{1}{2}$ meters. The depth of the reservoir is $5\frac{4}{5}$ meters. Calculate the mass of the water in this reservoir, if 1 dm^3 of the water in the reservoir has a mass of 1 kg.

9. A farmer harvested 1,502 kg of rice from $\frac{5}{7}$ ha of farmland. What quantity of rice can be harvested from 3.4 ha of such farmland under identical conditions?

10. There were 18 persons present during a meeting of a committee. This is nine-tenths of the total number of members the committee has. What is the total number of members which the committee has?

11. On the first two days of work, a tractor operator ploughed 18.75% of a farmland. The total area of the farmland is equal to 350 hectares. Calculate the area of the farmland which still remained for the tractor operator to plough?

12. Madina has both green and red balloons. The number of green balloons is 20; the number of red ones is 4 times more than the number of the green ones. What percent of all the balloons is the number of the red balloons?

13. In the 6th grade final exam in mathematics, 10 pupils got the mark of "100"; 25 pupils got the mark of "80"; and 15 pupils got the mark of "70". Find: the percent of the total number of the pupils who got the mark of "100"; the percent of the total number of the pupils who got the mark of "80"; and the percent of the total number of the pupils who got the mark of "70".

14. A kind of corn bread was found to contain 75% of corn meal. What quantity of corn bread was baked if 300 kg of corn meal was used in its preparation?

15. Five years from now, Kotee will be 3 times older than he is at present. How old is Kotee at present?

16. A tailor used 180 yards of cloth to sew 35 pairs of trousers and 30 shirts. For each pair of trousers he used 0.5 yard of cloth more than he used for each shirt. What is the quantity of cloth the tailor used on each pair of trousers and on each shirt?

17. The quantity of bananas in 3 kinjas is 165 kg (a kinja is a kind of country box usually made from fresh palm thatches and used mainly in the rural areas for carrying loads). The first kinja contains 52 kg. The second kinja contains 18 kg of bananas more than in the third. What quantities of bananas are contained in the second and third kinjas?

18. A worker can do a piece of work in 5 hours. Another worker can do the same work in 8 hours. How much of this work can they do in 1 hour, if they work together?

19. The quantity of coconut oil in one jar is 3.5 times more than that in a second jar. If 2.5 liters of the oil is poured from the first into the second jar, then the oil in both jars will become equal. What quantity of coconut oil is in each jar?

20. Two workers can do a piece of work together in 3 hours. In what time can one of the workers do this work alone, if the other worker can do it alone in 7 seven hours?

21. 7,680 kg of farina can be prepared from 9,600 kg of cassava. What quantity of cassava is required in order to prepare 560 kg of farina?

22. If 150 kg of sweet potatoes contain 120 kg of carbohydrates, then calculate the percent of carbohydrate contained in sweet potatoes.

23. In a certain grade of iron ore, for every 12 parts of iron there is 5 parts of mixtures (particles or components other than iron). What quantity of iron was smelted from this grade of iron ore if the quantity of mixtures was found to be 600 kg?

24. For the planting of 4.5 acres of a farmland, 18 tons of seed rice were required. What quantity of seed rice could be required for the planting of 2.5 acres of such a farmland?

25. An iron ball the volume of which is 9 cm^3 has a mass of 142.5 kg. What is the mass of a ball turned (or made) from an identical iron, if the volume of such ball is 4 cm^3?

26. 75 jars of fresh orange juice are prepared from 1,250 kg of oranges. How many such jars of fresh orange juice can be prepared from 500 kg of the same kind of oranges?

27. The actual distance between two villages is 350 km. What is the length (in centimeters) of a line representing this distance on a map drawn to a scale of 1:1,000,000?

28. The distance of 150 km between two rivers is represented on a map by a line segment which is 5 mm in length. Calculate the length of a line segment which would represent the distance of 1350 km on a map drawn to this scale.

29. The perimeter of a rectangle is 150 km. The length of the rectangle is 15 km more than its width. Calculate the sides and area of such a rectangle.

30. The base of a rectangular parallelepiped is a square length of whose side is 4 cm. If the volume of the rectangular parallelepiped is 32 cm^3, calculate its height and the area of its total surface.

Aptitude Test Five

1. What is the value of 3 raised to the third power?

2. What is the value of 2^5?

3. What is the cube root of 64?

4. What is the value of the expression $\sqrt[4]{10000}$?

5. The value of *n* raised to the third power is 125. What is *n*?

6. What is the equivalent in the decimal system of the Roman numeral MLMXLVIX?

7. What are the equivalents of the binary numbers 101101_2 and 1101.011_2 in the base-ten numeration system?

8. Calculate the simple interest and the amount on a bank deposit of $45,000 at the rate of 8% for $10\frac{1}{2}$ months.

9. Compare the two common fractions $\frac{6}{7}$ and $\frac{7}{8}$. Which is greater and by how much?

10. Expand the number 210 as the product of its prime factors.

11. Find the GCD and LCM of the numbers 60 and 280.

12. Find the GCD and LCM of the numbers 25, 40, and 70.

13. Find the GCD and LCM of 72, 84 and 96.

14. Round off the number 64.7538 to the nearest tenth.

15. Find the greatest common measure of the quantities:
    (a) 7.5 m and 1 m 5 dm; (b) 20 a and 600 m^2.

16. What is the least number of yards of cloth which must be in a roll of cloth so that it would be possible to sell it either by 5 yards or by 6 yards without remainder?

17. The perimeter of an isosceles triangle is 56.1 ft. The length of its base is 1.2 times less than each of its lateral side. Find the length of each side of this triangle.

18. Draw a triangle ABC with vertices in the following points on a coordinate plane: **A** (−4; −1.5), **B** (1; 2), **C** (3; −1.5). Find the area of this triangle.

19. Find the perimeter of triangle ABC as described above in problem number 18. Round off your answer to the nearest hundredth.

20. The length of a circumference is 31.4 cm. Find its radius and calculate the area of the circle enclosed by this circumference.

21. Draw a square MNOP with vertices in the following points on a coordinate plane: $\mathbf{M}(-1; -1)$, $\mathbf{N}(-1; 3)$, $\mathbf{0}(3; 3)$, $\mathbf{P}(3; -1)$. Calculate its perimeter and area.

22. A swimming pool can be filled with water by a first pipe for 3 hours 30 minutes; and a second pipe can fill the same pool for 5 hours 15 minutes. For what time the pool may be filled if both pipes are opened simultaneously?

23. The ratio of the mass of copper to the mass of zinc in brass is 1.5. Calculate the mass of copper in a piece of brass containing 9.45 kg of zinc.

24. The arithmetic mean of two numbers is equal to 48. One of the numbers is 4 times greater than the other. Find these two numbers.

25. Find the natural solution(s) to the equation $(y - 6) \cdot (y + 5) = 0$.

Solve the following equations:

26. $5x + 15 = 3 - 4x$.

27. $3\frac{1}{2}x + 1\frac{1}{4} = 1\frac{2}{3}x + 2\frac{1}{2}$.

28. Pewee can do a piece of work in 3 hours, and Teekonbla can do the same work in 4.5 hours. In what time can they do this work if the work together?

29.* Mr. Bangura deposited an amount of $12,250.00 in a bank paying an annual interest rate of 8%, compounded quarterly. Calculate the amount which the bank is supposed to pay to Mr. Bangura after 1 year.

30.* Mr. Dennis is told that the interest rate on his bank deposit is 8% per annum, compounded quarterly. Calculate the *equivalent effective annual rate* of interest on Mr. Dennis' bank deposit.

Answers to Problems and Exercises

Chapter One

1. (a) (i) $\dfrac{9}{100}$ = 0.09; (iii) $\dfrac{11}{200}$ = 0.055; (v) $\dfrac{7}{4}$ = 1.75; (b) (i) $\dfrac{1}{4}$ = 0.25; (iii) $\dfrac{61}{400}$ = 0.1525; (c) (ii) $\dfrac{2}{1}$ = 2; (iv) $\dfrac{2001}{1000}$ = 2.001; (d) (ii) $\dfrac{27}{25}$ = 1.08; (iv) $\dfrac{1}{2000}$ = 0.0005. **2.** (a) (i) 0.09 = 9 %; (iii) 1.08 = 108 %; (b) (i) 0.6 = 60 %; (iii) $\dfrac{2}{75} \approx 0.027 \approx 2.7$ %; (c) (i) 0.12 = 12 %; (iii) 0.9 = 90 %; (d) (i) 0.06 = 6 %; (iii) $\dfrac{19}{12} \approx 1.58 \approx 15.8$ %; (e) (i) 1.25 = 125 %; (iii) 5.2065 = 520.65 %. **3.** (a) $22.50; (c) $900; **4.** (a) Since 0.44 = 44 %, then 44 % = 0.44; (c) Since $\dfrac{7}{10}$ = 0.7 = 70 %, then $\dfrac{7}{10}$ = 70 %. **5.** (a) $1,182.64 – the compound interest on $8,447.40 at the rate of 14 % for 3 years; (b) $126 – the simple interest on $1,200 at the rate of 14 % for 9 months. **6.** 21 (kg) - the quantity of dry lemon grass that can be obtained from 60 kg of fresh lemon grass. **7.** 500 g = 0.5 kg - the mass of 1 natural orange juice bottle. **8.** five hours – time within which the ship and ferry boat will meet. **9.** 540 – number of components which still remain for him to manufacture. **10.** 84 – the arithmetic mean of his scores in the three subjects. **11.** Both Ali and Mohammed have equal number of purple balloons. **12.** 11.6928. **13.** 100 – the number in question. **14.** 300 (kg) – quantity of coconuts necessary to produce 180 kg of virgin coconut oil. **15.** 25 (kg) – the quantity of palm nuts necessary to prepare 2.5 kg of palm butter. **16.** 9.72 (kg) – the mass of dry lemon grass in 35 kg of fresh lemon grass. **17.** 11: 9 - the ratio of girls to boys in the 6th grade class. **18.** $600 - the total sum of money he paid for the 3 items. **19.** 20 (kg) – the quantity of dried bony that can be obtained from 60 kg of fresh bony. **20.** 100 %. **21.** 200 %. **22.** (a) 25 %; (c) 5 (times). **23.** (a) 400%; (b) 30%. **24.** (a) 0.25 times; (b) 0.6 times. **25.** (b) 1,200%; (c) 500%. **26.** (a) 20%; (b) 94.12% - the percent of coffee bag that was good. **27.** (a) 2.18%; (b) 9,300 %. **28.** 13.04%. **29.** 37.5% **30.** 16.67% - percent of those whose average score is 96; 41.67% - percent of those whose average score is 81; 33.33% - percent of those whose average score is 73; 8.33% - percent of those whose average score is 60; 55 – number of students making progress in their studies, that is achieving scores of 70 and above; 91.67% - percent of students making progress in their studies. **33.** 71.43 (kg) **35.** 100,000 (kg) **36.** 21.05% - percentage composition of banana in the cocktail; 31.58% - percentage composition of water in the cocktail; 2.11% - percentage composition of dry cream in the cocktail. **37.** (a) 1 : 5; (c) 2 : 1; (e) 10 : 1; (g) 5 : 1; **38.** (b) $x = \dfrac{3}{4}$; (d) $x = 1\dfrac{1}{3}$; (f) $x = 2.5$; (h) $x = 9$. **39.** 61:102 – the ratio of the light pole's height to the length of its shadow. **40.**

(a) The given ratio decreases by 4 times. **41.** The given ratio increases by $\frac{2}{3}$ times. **42.** The given ratio increases by 4 times. **43.** (a) The given ratio does not change; (b) The given ratio does not change. **44.** (a) 3 : 2; (c) 10 : 13; (e) 25 : 8; (f) 2 : 3. **45.** 1 : 5,000,000 – numerical scale to which the map is drawn. **46.** 7.5 cm **47.** 0.3 mm – length of the garden fence on the plan; 0.2 mm – width of the garden fence on the plan; 0.06 mm^2 – area of the garden fence on the plan. **48.** 768 (ft^2) **50.** (b) $x = 7$; (d) $x = \frac{3}{28} \approx 0.107$. **51.** $\frac{8}{15}$ (gallon) - the quantity of palm oil that can be produced from 10 kg of palm nuts. **52.** 17 (suits) – the number of suits that can be sewn from 85 yards. **53.** (b) $x = 34$; (d) $x = 9$. **54.** 19.5 hours **55.** 375 (days) **56.** 34 **57.** 250 **58.** 25 (kg) – the mass of corn meal necessary to bake 75 kg of the bread; 15 (kg) – the mass of eggs necessary to bake 75 kg of the bread; 5 (kg) – the mass of cream necessary to bake 75 kg of the bread. **59.** 450 (kg) – the mass of iron required to obtain 900 kg of the alloy; 300 (kg) – the mass of copper required to obtain 900 kg of the alloy; **60.** if the side a of a square is increased 4 times, then its area a^2 increases 16 times. **61.** If one of the sides of a rectangle is tripled (i.e. increased by 3), then the area of the rectangle increases three times. **62.** if the edge of a cube is increased 5 times, then its volume increases by 125 times. **63.** 6.25 (or 6 hours 15 minutes) **64.** 75 lb – the value of the 15-inch weight that could be suspended from the other arm of the lever to balance the 25-lb load. **65.** 2,400 m^2 **66.** 125,000 kg **67.** (a) (i) 25 % – percent of the total number of pages of the document she typed on the first day (given); (ii) 37.5 % – percent of the total number of pages of the document she typed on the second day; (iii) 37.5% - percent of the total number of pages of the document she typed on the third day; (b) 200 pages; (c) (i) 50 (pages) – the number of pages she typed on the first day (given); (ii) 75 (pages) – the number of pages she typed on the second day; (iii) 75 (pages) – the number of pages she typed on the third day; (d) 2 : 3 : 3 – the numeral proportion of the number of pages she typed on the first, second, and third day, respectively.

Chapter Two
2. $\angle RCO = 30^0$; $\angle TCO = 150^0$; **3.** 3 miles $\le$ CD $\le$ 13 miles; **4.** (a) True; (b) True; (c) True; (d) False; (e) True. **5.** $\angle RTX = 35^0$; $\angle RTZ = 125^0$; $\angle XTS = 110^0$. **6.** NP > MO **7.** P = 24 (cm); A = 21 (cm^2). **8.** One side is equal to 18 cm, and the other side is equal to 15 cm. **9.** CE = 6 (cm). **11.** $h = 4$ cm. **12.** $A_{real} = 1{,}250$ (m^2). **13.** The square has the greatest area. The perimeters of the given polygons are equal. **14.** 105,000 (m^2) – the total area of the rectangular field; 2,250 (m^2) – the area of the portion of the highway covered by the field; 102,750 (m^2) – the remaining area of the field (apart from the area of its portion through which the highway passes). **15.** In increasing the length of its side by 4 times, the area of a square increases accordingly by 16 times. **16.** In order that the area of a square is decreased by 9 times, it is necessary to decrease its side

by 3 times. **17.** The length of one side of the rectangle is 60 cm; the length of the other side of the rectangle is 75 cm; 4500 (cm) – the area of the rectangle. **18.** The perimeter of the rectangle is less than that of the parallelogram. **19.** 80.5 (cm^2) – the area of the trapezoid. **20.** 17.25 cm^2 – the total area of the figure. **21.** 341.72 cm^2 – the area of the trapezoid. **22.** 33,750 kg – the quantity of kilograms of seeds required to sow the farmland in the form of a trapezoid. **23.** 86,875 m^2 (or 8.6875 hectares) – the actual area of the given plot of land. **25.** 324 cm^2 – area of the lateral surface of the rectangular parallelepiped; 478 cm^2 – area of the total surface of the rectangular parallelepiped; 693 cm^3 – the volume of the rectangular parallelepiped; **26.** 2.25 cm^2 – the area of the triangular base of the prism; 30.4 cm^2 – the area of the total surface of the triangular prism. **27.** 84 (cm^2) – the area of the lateral surface of the triangular prism; 96 (cm^2) – area of the total surface of prism; 42 (cm^3) – the volume of the prism. **28.** 556 (cm^2) – area of the total surface of the prism; 600 (cm^3) – the volume of the prism. **30.** 54 (cm^2) – the area of the total surface of the quadrilateral pyramid. **31.** 75 (cm^3) – the volume of the pyramid.

Chapter Three
1. 31.4 (cm) – the circumference of the circle; d = 10 cm, its diameter. **2.** 7 (cm) – the diameter of the circumference. **4.** The diameter is the length of the greatest chord of a circle or circumference. **5.** Radius = 20 mm (when point P lies within the circumference); Radius = 5 mm (when point P lies outside the circumference). **6.** 40,054 (km) – the perimeter (or length) of the Earth's equator. **7.** 6,786.19 (km) – diameter of the planet Mars. **8.** 9.42 (km) – length of the circumference of the horizon. **9.** 40.82 (cm) – the distance which the end of the minute hand of the wall clock can go for 12 hours. **10.** 785 (cm) – the distance at which the load is lifted for ten rotations (or turns) of the pulley. **11.** 26 (times) – the number of times is it necessary to turn or rotate the handle of the shaft in order to draw out a bucket of water from the well. **12.** 24 (cm) – as much in length the radius of the circumference would increase, if the length of the circumference was increased by 1.5 m. **13.** 7.85 (cm) – as much the length of the circumference would decrease as a result of the decrease in its diameter. **14.** 66.91 (hours) – the time for which it is possible to fly in an airplane around the Earth along the equator, moving with the speed indicated. **15.** 2,826 (cm^2) – the area of the circle. **16.** 7.31 (cm^2) – the area of the ring of a section of the pipe. **17.** 8.86 (cm^2) – the area of the shaded figure. **18.** 17.5 (miles^2) – the area of the complex of the football stadium. **19.** 3 (times) – as much the length of the larger circumference is greater than that of the smaller; 9 (times) – as much the area of the larger circle is greater than the area of the smaller one. **20.** 27.997 (cm^2) – the area of the circle (rounded off to the nearest thousandth). **21.** 251.6 (in^2) – the total area of the gasket (rounded off to the nearest tenth). **22.** 1,962.5 (cm^2) – the area of the largest possible circle that is cuttable from the

rectangular piece of paper; 56.39 % - the percentage of the rectangular piece of paper which is wasted as a result of cutting the circle from it. **23.** 45.2548 ≈ 45.3 (cm) – diameter of the new water pipe. **24.** The new dimensions are: 31.7925 (mm^2) – the new cross-sectional area of the aluminum wire (which is two times less than that of the original); 6.36 (mm) – the new (reduced) diameter of the stretched aluminum wire. **25.** 339.12 (cm^2) – the area of the cylinder's lateral surface; 254.34 (cm^2) – the area of each of the cylinder's circular bases; 847.8 (cm^2) – the area of the cylinder's total surface; 1,526.04 (cm^3) – the cylinder's volume. **26.** 2,355 (cm^2) – the area of the total surface of the cone; 5887.5 (cm^3) – the volume of the cone. **27.** 9156.24 (cm^2) – the area of the sphere; 82,406.16 (cm^3) – the volume of the sphere. **28.** 4.63 (times) – the number of times the mass of the larger ball-bearing is greater than that of the smaller. **29.** 5.832 (times) – the number of times the volume of the larger pineapple is greater than that of the smaller; it is more profitable and advantageous to buy the larger pineapple. **30.** 13,377.91 (in^3) – the volume of the sphere. **31.** 510,926,783.04 (km^2) – the area of the surface of the Earth; 40,053.84 (km) – the length of the Earth's equatorial circumference.

Chapter Four: Aptitude Tests

Aptitude Test One.

1. 15.6; **2.** $6\frac{34}{75}$; **3.** $-8\frac{1}{9}$; **4.** 7.29 ; **5.** $-\frac{15}{49}$; **6.** 22.575; **7.** 48; **8.** $\frac{1}{2}$; **9.** −13.2; **10.** 51; **11.** $4\frac{13}{128}$; **12.** −50.5; **13.** −10; **14.** $\frac{49}{81}$; **15.** $3\frac{7}{36}$; **16.** $-$ 46.55; **17.** 7.75; **18.** $-\frac{104}{675}$; **19.** $131\frac{1}{6}$; **20.** 19.875; **21.** $\frac{197}{450}$; **22.** $3\frac{5}{6}x - 4\frac{7}{9}$; **23.** $35x - 42x^2$; **24.** $-3a +\frac{3}{8}$; **25.** $-14a + 12ab -21b$; **26.** $- 27x^2 + 36x$); **27.** $-25\frac{3}{4}a - 2\frac{2}{5}$; **28.** −21; **29.** −25.875; **30.** − 5.9;

Aptitude Test Two.

1. 1.5; **2.** 2.9; **3.** $- 8\frac{7}{25}$; **4.** − 39; **5.** 44.1; **6.** $\frac{239}{975}$; **7.** 2(a+1); **8.** −3(x + 7y) +14; **9.** 11(0.2a − 0.7b + 0.9); **10.** $- (1\frac{1}{7}x + \frac{1}{3}y + 1\frac{2}{3})$; **11.** 84; **12.** $\frac{35}{36}$; **13.** $x = 3\frac{25}{42}$; **14.** a = −22.35; **15.** x = −14; **16.** $1\frac{1}{14}$; **17.** y = 10.4; **18.** $8\frac{41}{53}$; **19.** $b = \frac{3}{5}$; **20.** x = $4\frac{1}{6}$; **21.** $x = \frac{27}{40}$; **22.** $y = 133\frac{1}{3}$ or $y = -133\frac{1}{3}$; **23.** x = 0.15 or x = − 0.15; **24.** x = $\frac{7}{216}$ or $x = -\frac{7}{216}$; **25.** a = 29 or a = − 29; **26.** a = 7; **27.** x = 18; **28.** y = 4$\frac{10}{37}$; **29.** $a = 3\frac{3}{53}$; **30.** $y = \frac{8}{67}$;

Aptitude Test Three

1. $k = -4.5$; **2.** $n = -\dfrac{5}{24}$; **3.** $-1\dfrac{3}{16}$; **4.** 1.5; **5.** 0.32; **6.** $(-\dfrac{5}{6})^2$ is greater than $(-\dfrac{4}{5})^3$ by $1\dfrac{929}{4500}$; **7.** $(-0.32)^2$ is greater than $(-3.2)^3$ by 32.8704; **8.** $\left|(-\dfrac{3}{4})^2\right|$ is $8\dfrac{101}{128}$ times greater than $\left|(-\dfrac{2}{5})^3\right|$; **9.** $\left|(-\dfrac{1}{2})^2\right|$ is $\dfrac{27}{32}$ times greater than $\left|(-\dfrac{2}{3})^3\right|$; **10.** $-33\dfrac{17}{30}$; **11.** 6.5; **12.** -13.29; **13.** 3.8; **14.** The difference of $2\dfrac{3}{4}$ and $\dfrac{7}{8}$ is less than the quotient of $-15\dfrac{1}{2}$ and $-3\dfrac{3}{4}$ by $2\dfrac{31}{120}$; **15.** $5\dfrac{41}{45}$; **16.** $-87\dfrac{4}{5}$; **17.** \$18,000; **18.** 6.375 miles; **19.** 400; **20.** 1,200; **21.** 10%; **22.** 4%; **23.** 12.5%; **24.** $-3\dfrac{1}{3}$; **25.** $x = \dfrac{2}{3}$; **26.** $y = \dfrac{20}{23}$; **27.** -81; **28.** -14.2; **29.** $x = -1\dfrac{5}{26}$; **30.** $a = \dfrac{35}{624}$;

Aptitude Test Four.

1. $d = \dfrac{3}{22}$; **2.** $d = \dfrac{35}{48}$; **3.** 72 mph – the speed of the bus; **4.** 3.15 (kg) – the mass of $\dfrac{3}{8}$ dm^3 of the iron; **5.** 1,084 $\dfrac{11}{46}$ (tons) – the mass of the layer of the rock having the given dimensions; **6.** 160 km – distance the taxi cab would travel for $2\dfrac{2}{3}$ hours under the same rate of motion; **7.** $2\dfrac{7}{15}$ (hours) = 2 hours 28 minutes – the time Mr. Sumo spent on his journey between his farm and the village; **8.** 863.04 (tons) – the mass of the water in the reservoir; **9.** 7,149.52 (kg) – the quantity of rice that can be harvested from 3.4 ha of such farmland under identical conditions; **10.** 20 (persons) – the total number of members which the committee has; **11.** 284.375 (ha) – the area of the farmland which still remained for the tractor operator to plough; **12.** 80% – the percent of all the balloons which is equal to the number of red ones; **13.** 20% – the percent of the total number of the pupils who got the mark of "100"; 50% – the percent of the total number of the pupils who got the mark of "80"; 30% – the percent of the total number of the pupils who got the mark of "70"; **14.** 400 (kg) – the quantity of corn bread that was baked from 300 kg of corn meal; **15.** 2.5 (years) – the age of Kotee at present; **16.** 2.5 (yards) – the quantity of cloth the tailor used on each shirt; 3 (yards) – the quantity of cloth the tailor used on each pair of trousers; **17.** 65.5 kg – the quantity of bananas in the second kinja; **18.** $\dfrac{13}{40}$ – the fractional part of the work which the two workers can do for 1 hour, if they work together; **19.** 2 (liters) − the quantity of coconut oil in the second jar; 7 (liters) − the quantity of coconut oil in the first jar; **20.** $5\frac{1}{4}$ (hours) = 5 hours 15 minutes − the time in which the first worker alone can do the work; **21.** 700 (kg) − the quantity of cassava required in order to prepare 560 kg of farina; **22.** 80 % − the percent of carbohydrate contained in sweet potatoes; **23.** 1,440 (kg) − the quantity of iron ore smelted from this grade of iron ore containing 600 kg of mixtures; **24.** 10 (tons) − the quantity of seed rice required to plant 2.5 acres of the farmland; **25.** $63\frac{1}{3}$ (kg) − the mass of the iron ball having the volume of 4 cm³; **26.** 30 (jars) − the number of   jars of fresh orange juice which can be prepared from 500 kg of the same kind of oranges; **27.** 35 cm − the length of the line representing the actual distance between the two villages on a map drawn to the given scale; **28.** 45 (mm) – the length of a line segment which would represent the distance of 1350 km on the map drawn to such scale; **29.** 30 (km) − the width of the rectangle; 45 (km) – the length of the rectangle; 1350 (km²) – the area of the rectangle; **30.** 2 (cm) – the height of the rectangular parallelepiped; 64 (cm²) − the area of the total surface of the rectangular parallelepiped;

Aptitude Test Five.

1. 27; **2.** 32; **3.** 4; **4.** 10; **5.** $n = 5$; **6.** 2004; **7.** $13\frac{3}{8}$; **8.** $3,150 – the simple interest on the bank deposit; $48,150 – the amount; **9.** $\frac{7}{8}$ is greater than $\frac{6}{7}$ by $\frac{1}{56}$; **10.** $210 = 2 \cdot 3 \cdot 5 \cdot 7$; **11.** GCD (60, 280) = 20; LCM (60, 280) = 840; **12.** GCD (25, 40, 70) = 5; LCM (25, 40, 70) = 1400; **13.** GCD (72, 48, 96) = 12; LCM (72, 48, 96) = 2016; **14.** 64.7538 ≈ 64.8; **15.** (a) GCD (75, 15) = 15; (b) GCD (20, 6) = 2; **16.** LCM (5, 6) = 30; **17.** 16.5 (ft) − the length of the base of the isosceles triangle; 19.8 (ft) − the length of each of its lateral sides; **18.** 12.25 (square units) – the area of triangle ABC; **19.** 17.13 (units) − the perimeter of triangle ABC as described in problem number 18; **20.** 5 (cm) – radius of the circumference; 78.5 (cm²) – the area of the circle enclosed by the circumference; **21.** 16 (units) – the perimeter of the square MNOP; 16 (square units) – the area of the square MNOP; **22.** $2\frac{1}{10}$ hours = 2 hours six minutes − the time for which the pool may be filled if both   pipes are opened simultaneously; **23.** 14.175 (kg) – the mass of copper in a piece of brass containing 9.45 kg of zinc; **24.** 19.2 – the smaller number; 76.8 – the greater number; **25.** $y = 6$; **26.** $x = -1\frac{1}{3}$; **27.** $x = \frac{15}{22}$; **28.** $1\frac{4}{5}$ hour = 1 hour 48 minutes − the time in which Peewee and Teekonbla can do the work if the work together; **29.** $13,259.79 − the amount which the bank is supposed to pay to Mr. Bangura after 1 year; **30.** 8.24% − the equivalent effective annual rate of interest on Mr. Dennis' bank deposit.

APPENDIX

I. Tables of Weights and Measures

The Metric System

A. Linear Measure (or Units of Length)

1 mm = 0.03937 in
10 mm = 1 cm = 0.3937 in
10 cm = 1 dm
10 dm = 100 cm = 1 m = 39.37 in =
3.2808 ft

1,000 m = 1 km = 0.621 mile =
3,280.8 ft
1 dm = 100 mm
1 m = 1,000 mm
1 km = 100,000 cm

B. Square Measure (or Units of Area)

$1 \text{ mm}^2 = 0.00155 \text{ in}^2$
$100 \text{ mm}^2 = 1 \text{ cm}^2 = 0.15499 \text{ in}^2$
$100 \text{ cm}^2 = 1 \text{ dm}^2$
$100 \text{ dm}^2 = 1 \text{ m}^2 = 1,549.9 \text{ in}^2 =$
1.196 yd^2
$1,000,000 \text{ m}^2 = 1 \text{ km}^2$

$100 \text{ m}^2 = 1 \text{ are (a)} = 119.6 \text{ yd}^2$
100 a = 1 hectare (ha)
100 square hectometers (hm^2) = 100
ha = $1 \text{ km}^2 = 0.386 \text{ mile}^2 = 247.1$
acres
$1 \text{ acre} = 4,046.86 \text{ m}^2 = 4,840 \text{ yd}^2$

C. Volume Measure (or Units of Volume)

$1,000 \text{ mm}^3 = 1 \text{ cm}^3 = 0.06102 \text{ in}^3$
$1,000 \text{ cm}^3 = 1 \text{ dm}^3$ (1 liter) = 61.023
$\text{in}^3 = 0.0353 \text{ ft}^3$
10 liters = 1 decaliter
100 liters = 1 hectoliter

$1,000 \text{ dm}^3 = 1 \text{ m}^3 = 35.314 \text{ ft}^3 =$
1.308 yd^3
$1,000,000 \text{ cm}^3 = 1 \text{ m}^3$
$1,000,000,000 \text{ m}^3 = 1 \text{ km}^3$

D. Weight/Mass Measure (Units of Weight or Mass)

10 decigrams (dg) = 1 gram (g) =
15.432 grains = 0.035274 ounce
(avdp);
1,000 mg = 1 g
1,000 g = 1 kg

10 hectograms (hg) = 1 kg = 2.2046
pounds
100 kg = 1 centner (cnt)
10 cnt = 1,000 kg = 1 ton (T)
10 quintals = 1 metric ton = 2204.6
pounds

E. Land Measure

100 centiares = 1 are (a) = 119.6 yd^2 100 a = 1 hectare (ha) = 2.471 acres

100 ha = 1 km^2 = 0.386 mile^2 =   0.386 mile^2 = 247.1 acres

The British (or Imperial) System

A. Linear Measure (or Units of Length)

1 mil = 0.001 in = 0.0254 mm
1 in = 1,000 mils = 2.54 cm
12 in = 1 ft = 0.3048 m
3 ft = 1 yd = 0.9144 m
$5\frac{1}{2}$ yd = $16\frac{1}{2}$ ft = 1 rod (or pole) =
5.029 m

40 rods (or poles) = 1 furlong =
201.168 m
8 furlongs = 1760 yd = 5280 ft = 1
mile = 1.6093 km

B. Square Measure (or Units of Area)

1 in^2 = 6.452 cm^2
144 in^2 = 1 ft^2 = 929.03 cm^2
9 ft^2 = 1 yd^2 = 0.8361 m^2
$30\frac{1}{4}$ yd^2 = 1 rod^2 (or pole^2) = 25.292
m^2

160 rod^2 = 4840 yd^2 = 43,560 ft^2 = 1
acre = 0.4047 ha
640 acres = 1 mile^2 = 259.00 ha =
2.590 km^2

C. Cubic Measure

1 in^3 = 16.387 cm^3
1728 in^3 = 1 ft^3 = 0.0283 m^3

27 ft^3 = 1 yd^3 = 0.7646 m^3

II. Useful Equivalents

1 mile = 1760 yards = 1.609
kilometers (km)
1 km = 0.621 mile
1 inch = 2.54 centimeters (cm)
1 cm = 0.3937 inch
1 meter (m) = 39.37 inches = 3.28
feet (ft) = 1.09 yard
1 yard = 0.9144 meter (m)
1 kg = 2.204 pounds (lb)

1 pound = 0.453 kg= 16 ounce (oz)
1 oz = 28.35 grams (g)
1 hour = 60 minutes = 3,600 seconds
100 m^2 (square meters) = 1 are (a)
100 ares = 1 hectare
1 hectare = 10,000 m^2 = 2.471 acres
1 m^2 = 100 dm^2 = 10,000 cm^2
1 km^2 = 1,000,000 m^2
I gallon (gal) = 3.79 liters

III. List of Abbreviations and Symbols

a = are(s)
ans = answer
avdp =avoirdupois
kg = kilogram
mm = millimeter
m = meter
m^2 = square meter(s)
m^3 = cubic meter(s)
mi = mile(s)
mi^2 = square mile(s)
sq= square
g = gram
 lb = pound
oz = ounce
cm = centimeter
hr = hour
min = minute
sec = second
mph = miles per hour
kph = kilometer per hour

yd = yard(s)
yd^2 = square yard(s)
yd^3 = cubic yard(s)
km = kilometer
$\in$ = belongs to (is a member of)
$\angle$ = angle
$\perp$ = perpendicular
$\parallel$ = parallel
$\triangle$ = triangle
$\leq$ = less than or equal to
$\geq$ = greater than or equal to
$\approx$ = almost equal to
$>$ = greater than
$<$ = less than
^0C = degree (s) Celsius
e.g. = for example
esp. = especially

ha = hectare(s)
T = ton(s)
cnt = centner(s)
dm = decimeter
etc. = et cetera
ft = foot (feet)
ft^2 = square foot (feet)
ft^3 = cubic foot (feet)
i.e. = that is
in = inch(es)
in^2 = square inch(es)
in^3 = cubic inch(es)
pt= point(s)
dg = decigram
hm = hectometer
hm^2 = square hectometer
hm^3 = cubic hectometer
hg = hectogram

IV. The Greek Alphabet

Capital letter	Small letter	Name of symbol	Associated meaning or value
A	α	Alpha	The first
B	β	Beta	The second
Γ	γ	Gamma	
Δ	δ	Delta	
E	ε	Epsilon	
Z	ζ	Zeta	
H	η	Eta	
Θ	θ	Theta	
I	ι	Iota	The smallest part

K	κ	Kappa	
Λ	λ	Lamda	
M	μ	Mu	
N	ν	Nu	
Ξ	ξ	Xi	
O	o	Omicron	
Π	π	Pi	≈ 3.14
P	ρ	Rho	Specific resistance
Σ	σ	Sigma	Summation
T	τ	Tau (Tao)	
Y	υ	Upsilon	
Φ	φ	Phi	
X	χ	Chi (Khi)	
Ψ	ψ	Psi	
Ω	ω	Omega	The last

V. A Table of Common Polygons and Polyhedrons and Their Formulae

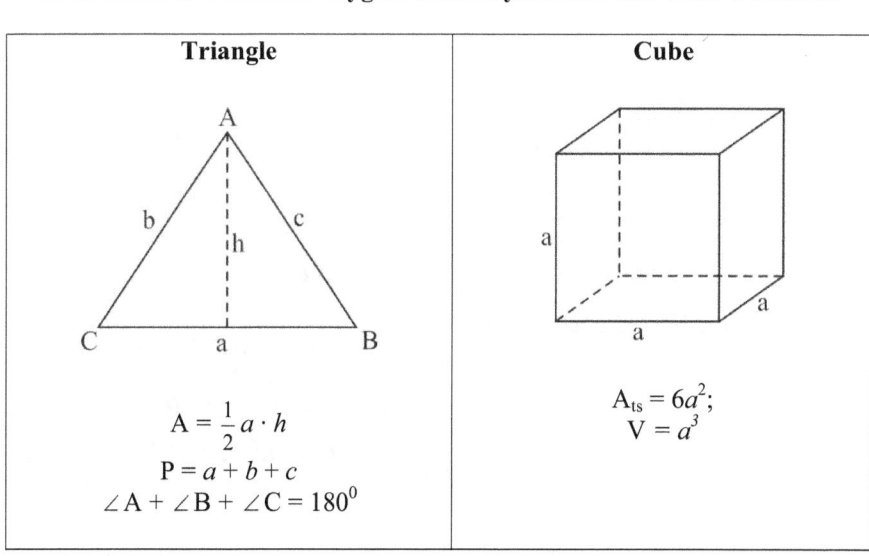

Triangle	Cube

$$A = \frac{1}{2}a \cdot h$$

$$P = a + b + c$$

$$\angle A + \angle B + \angle C = 180^0$$

$$A_{ts} = 6a^2;$$

$$V = a^3$$

Right Triangle	Triangular Prism
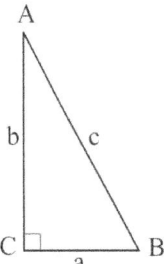	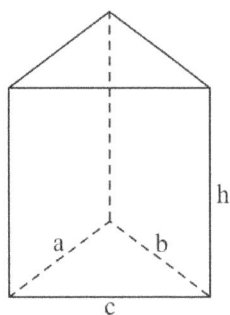
$A = \dfrac{1}{2} a \cdot b$ $P = a + b + c$ $\angle A + \angle B = 90^0$ $\angle C = 90^0$ $c^2 = a^2 + b^2$ $c = \sqrt{a^2 + b^2}$	$A_{ls} = (a + b + c) \cdot h = P_{base} \cdot h;$ $A_{base} = \dfrac{1}{2} (c \cdot h_b);$ h_b – height of the triangular base; $A_{ts} = A_{ls} + 2A_{base};$ $V = A_{base} \cdot h$
Rectangle	Triangular Pyramid
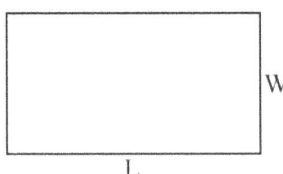	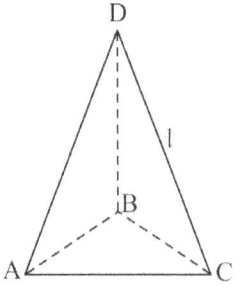
$A = l \cdot w$ $P = 2(l + w)$	$\mathbf{A_{ls}} = A_1 + A_2 + A_3,$ where A_1, A_2, A_3 – areas of lateral triangles (ADC, ADB, and BDC). $A_1 = \dfrac{1}{2} AC \cdot h_1;$ $A_2 = \dfrac{1}{2} AB \cdot h_2;$ $A_3 = \dfrac{1}{2} BC \cdot h_3;$ h_1, h_2, h_3 – heights of lateral triangles ADC, ADB, and BDC; $A_{base} = \dfrac{1}{2} (AC \cdot h_b);$

	$A_{ts} = A_{ls} + A_{base};$ $V = \dfrac{1}{3} A_{base} \cdot h$
Square	**Circle**
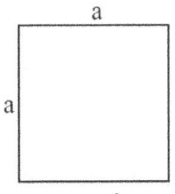 $A = a^2$ $P = 4a$	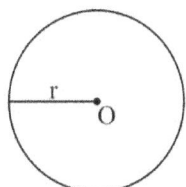 $\mathbf{c} = \pi d = 2\pi r;$ $A_{circle} = \pi r^2.$
Parallelogram	**Cylinder**
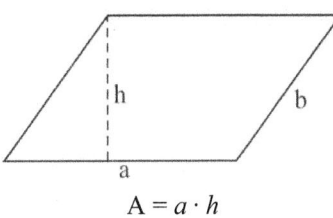 $A = a \cdot h$ $P = 2(a + b)$	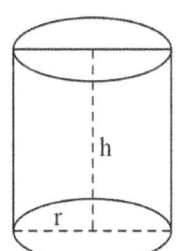 $A_{ls} = c \cdot h = 2\pi r h;$ $A_{base} = \pi r^2;$ $A_{ts} = A_{ls} + 2 A_{base};$ $V = A_{base} \cdot h = \pi r^2 h;$
Trapezoid	**Cone**
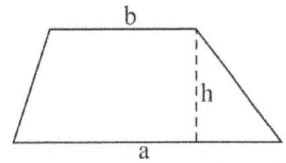 $A = \dfrac{1}{2}(a + b) \cdot h = \left(\dfrac{a+b}{2}\right) \cdot h$	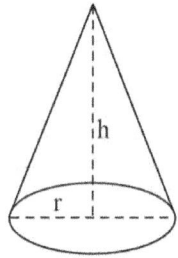 $A_{ls} = \dfrac{1}{2} lc = \pi r l; \ A_{base} = \pi r^2;$ $A_{ts} = A_{base} + A_{ls} = \pi r^2 + \pi r l =$

	$= \pi r\,(r + l)$; $V = \dfrac{1}{3}\,A_{base} \cdot h = \dfrac{1}{3}\,\pi r^2 h.$
Rectangular Parallelepiped 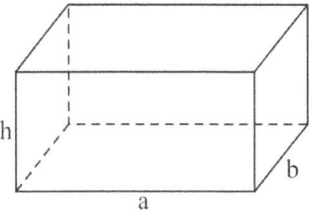 $A_{ls} = 2(a + b) \cdot h = P_{base} \cdot h$; $A_{base} = a \cdot b$; $A_{ts} = A_{ls} + 2A_{base}$ $V = a \cdot b \cdot h = l \cdot w \cdot h$; where $a = l$, and $b = w$.	**Sphere** 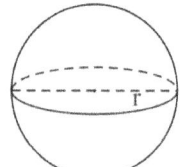 $A = 4A_{circle} = 4\,\pi r^2$; $V = \dfrac{4}{3}\,\pi r^3$;

<u>Legend</u>

a, b, c – sides
l = length
w = width
h = height
r = radius
V = volume
A = area

A_{ls} = area of the lateral surface
A_{base} = area of the base
A_{ts} = area of the total surface
P = perimeter

P_{base} = perimeter of the base
$c = 2\,\pi r$ – length of circumference

ABOUT THE AUTHOR

J. Nyenetu Jarkloh is a Liberian-born graduate professional communications engineer, senior corporate manager and administrator, educator, author and publisher, corporate communications specialist, policy analyst, and a staunch advocate for human rights.

In Monrovia (Liberia), he attended the Amanda Caphart Elementary, Boatswain Junior High, Government Junior High, W. V. S. Tubman High, before enrolling at the University of Liberia (UL) in the College of Business. In the wake of Liberia's 1980 coup, based on excellent performance in his studies at the UL, he was granted a bilateral scholarship to pursue advanced studies in the former Soviet Union, where he earned a master's degree in radio engineering from the Odessa State Polytechnic University (Ukraine)

Upon return to his native country, Mr. Jarkloh had worked as a senior instructor in the Electronics Engineering Department at the W. V. S. Tubman Technical College (now Tubman Technical University) in Harper (Maryland County, Liberia), a part-time lecturer in mathematics (Numerical Analysis) at the University of Liberia, and also as Manager of Planning, Research & Development Department of the Liberia Telecommunications Corporation (L.T.C.) (now Liberia Telecommunications Authority). He also holds post-graduate professional certificates in telecommunications planning and management from the Satellite Transmission Systems Inc (Long Island, N.Y.), INTELSAT, and the United States Telecommunications Training Institute (Washington, D. C., U.S.A).

As a Fellow of the International Telecommunications Union (ITU), Mr. Jarkloh had actively participated in the ITU's GMDSS project, which called for elaboration of national master plans from English-speaking African Countries for the Development of Maritime Radiocommunications services. He elaborated his country's "National Master Plan for the Development of Maritime Radiocommunication Services in Liberia", which he defended during the final workshop of said project at the International Center for Theoretical Physics (ICTP) in Trieste (Italy). Due to the Liberian Civil Crisis, however, he was evacuated. He presently resides with his family in Ukraine.

www.ingramcontent.com/pod-product-compliance
Lightning Source LLC
Chambersburg PA
CBHW051323170526
45166CB00002B/667